Marine Microbial Diversity as a Source of Bioactive Natural Products

Marine Microbial Diversity as a Source of Bioactive Natural Products

Special Issue Editor

Didier Stien

MDPI • Basel • Beijing • Wuhan • Barcelona • Belgrade

Special Issue Editor
Didier Stien
Sorbonne Université, CNRS,
Laboratoire de Biodiversité et
Biotechnologie Microbienne,
USR3579, Observatoire
Océanologique
France

Editorial Office
MDPI
St. Alban-Anlage 66
4052 Basel, Switzerland

This is a reprint of articles from the Special Issue published online in the open access journal *Marine Drugs* (ISSN 1660-3397) from 2018 to 2020 (available at: https://www.mdpi.com/journal/marinedrugs/special_issues/Marine_Microbial_Diversity).

For citation purposes, cite each article independently as indicated on the article page online and as indicated below:

LastName, A.A.; LastName, B.B.; LastName, C.C. Article Title. *Journal Name* **Year**, *Article Number*, Page Range.

ISBN 978-3-03936-182-3 (Hbk)
ISBN 978-3-03936-183-0 (PDF)

Cover image courtesy of Gwenaël Piganeau.

Contents

About the Special Issue Editor

Didier Stien is a CNRS Senior Scientist in Chemistry, working at the Banyuls Oceanological Observatory, Sorbonne University, France. He studied chemistry at Aix-Marseille University, France, where he obtained a Ph.D. in 1997. After a post-doctoral stay in Weinreb's group at Pennsylvania State University, USA, he was hired as a research fellow at CNRS in 1999. He worked at the French Guyana CNRS research center starting in 2003. He has developed natural product research inspired by the observation of networks of species, including bio-guided isolation and structure elucidation, chemical profiling, metabolomic studies of active natural or synthetic mixtures, and chemical ecology. At Sorbonne University since 2013, he is now leading research on the search for innovative antibacterial agents from marine microorganisms. He studies bacterial and fungal symbionts or pathogens of marine organisms such as plants, algae, or sponges in search of innovative active natural products. He is interested in the toxicity and fate of emerging pollutants as well. This includes the search for toxicity and stress markers in marine organisms and the search for biotransformation products of the tested pollutants. To date, he has published more than 100 peer-reviewed papers.

Editorial

Marine Microbial Diversity as a Source of Bioactive Natural Products

Didier Stien

Laboratoire de Biodiversité et Biotechnologie Microbiennes, Sorbonne Université, CNRS, LBBM, Observatoire Océanologique, 66650 Banyuls-sur-Mer, France; didier.stien@cnrs.fr

Received: 9 April 2020; Accepted: 14 April 2020; Published: 16 April 2020

Some 3.5 billion years ago, microorganisms were the first to colonize Earth. They have gradually evolved, within intricate systems of microbial and macroscopic species, to occupy virtually all the available niches on the planet. Genetic drift and natural selection have molded the phenotypic expression of a trillion different microbial species [1], shaping their metabolisms and offering ever-more advantageous abilities to expand, with the production of new metabolites being one key to the fitness and evolutionary success of new species [2].

The rate of discovery of new natural products of microbial origin has increased significantly since the 1970s with the advent of modern methods of purification and structural determination, and later, with holistic approaches to chemical analysis and genetic information-based exploration of natural chemodiversity [3–6]. With these tools in hand and the exploration of innovative and ecologically relevant sources of microorganisms, it is possible to rapidly expand the exploration of chemodiversity on a global scale, and eventually isolate innovative compounds. It should be noted that the enthusiasm to discover new scaffolds is not necessarily relevant in the context of the search for active or useful natural products. Structural modifications that do not affect the core scaffold of a metabolite may provide a significant benefit that can be developed for a therapeutic use. A notable example is in the immense diversity of terpenes built on a limited number of scaffolds. Terpenes have different functional roles—and therefore different biological activities—and have largely contributed to the diversification of higher plants [2,7].

Today, marine microbes can still be regarded as a relatively underappreciated potential source of active compounds yet to be discovered. Some microorganisms can live freely in the water column, while many others live in association. For example, sponges host microorganisms that contribute to their holometabolome, while many coral species host microalgae and symbiotic prokaryotes. Furthermore, many microorganisms contribute to the construction of interspecific biofilms covering biotic and abiotic surfaces. While recent advances in natural-product science now provide access to an unknown dimension of chemodiversity, metabolic engineering will eventually allow the production of active marine compounds in amounts necessary for pharmaceutical development.

The Special Issue "Marine Microbial Diversity as a Source of Bioactive Natural Products" was aimed at collecting papers with up-to-date information regarding the characterization of marine microbes' metabolic diversity and the evaluation of the therapeutic potential of marine microbes' metabolites. The interest of the Special Issue was also to show that the exploration of underrated reservoirs of marine microorganisms can lead to the discovery of valuable secondary metabolites. In total, 10 articles were accepted and included in the Special Issue.

Most of the articles in this special issue deal with marine fungi, biological and chemical diversity, and their active metabolites. This may be a sign that marine fungi have been under studied to date, and are perceived by many researchers as an important source of discovery in this field [8]. Focusing on marine *Arthrinium* spp. isolates, Heo et al. conducted a phylogenetic analysis based on internal transcribed spacers, nuclear large subunit rDNA, β-tubulin, and translation elongation factor region sequences [9]. The 28 analyzed strains were obtained by cultivation of seaweed tissues (including several *Sargassum fulvellum* individuals) and *Arctoscopus japonicus* egg masses. It was found that the 28 isolates were in fact 15 species, 11 of which being new to Science. Most of the fungal extracts exhibited radical-scavenging activity, and some showed antifungal activity, tyrosinase inhibition, and quorum sensing inhibition. Interestingly, three species were found in both *S. fulvellum* and *A. japonicus* egg masses, perhaps because these fish usually lay eggs on *S. fulvellum*. The known compound gentisyl alcohol was found to be responsible for the radical-scavenging activity of two of the *Arthrinium* extracts. A fungal diversity analysis was also conducted on decommissioned salterns and neighboring mudflats in the Yellow Sea of South Korea [10]. Comparative fungal community analysis showed that the salterns that had been abandoned for more than 35 years had recovered to mudflats. The Yongyudo saltern was abandoned less than one year before the analysis and had not recovered. Its fungal community was more diverse and was dominated by Entorrhizomycota, while Chytridiomycota and Mortierellomycota dominated elsewhere. It has been hypothesized that Entorrhizomycota may include plant pathogenic fungi and that the dominance of this phylum may originate from the occurrence of their host plants at the initial stage of ecological succession after the saltern was abandoned. It should be mentioned that a large number of fungi could not be identified, indicating a lack of DNA-based phylogenetic information on marine fungi. Eventually, 53 fungal strains were isolated from the different sampling locations. The cultivable fungi were not necessarily the main taxa from the community analyses. A total of 18 isolates were possibly new species, and the authors provide the antioxidant, antifungal, tyrosinase inhibition, and quorum quenching activity of all isolates.

Fungal spores are easily dispersed and are detectable in community analyses. As a result, not all fungi detected in marine environments by rDNA-based analysis are genuine marine species. The challenging question of the isolation of marine fungi sensu stricto is key to the future exploration of the chemical diversity of marine fungi, and possibly the exploitation of their secondary metabolites. A best-practice guide for the isolation of marine fungi from different matrixes and their conservation is presented by Overy et al. [11]. Generalist, osmotolerant/halotolerant genera such as *Aspergillus* and *Penicillium* are highly cited in the literature but taxonomic groups composed predominantly of marine fungi sensu stricto such as Halosphaeriaceae, Torpedosporales, and Lulworthiales have been little studied for now. It should also be noted that mycologists have been studying marine fungi for decades, and that currently cryopreserved marine fungi may represent an easily accessible source for the exploration of their chemical diversity and the isolation of active secondary metabolites.

One of the sources frequently explored in the search for marine fungi are sedimentary deposits. Examination of the full genome of the marine-derived fungus *Penicillium brasilianum* HBU-136, isolated from the Bohai sea in China, highlighted that the strain harbors multiple non-ribosomal peptide synthetase (NRPS) biosynthetic gene clusters (BGC), including one showing high similarity to the BGC of fumitremorgins A-C in the fungus *Aspergillus fumigatus* [12]. The strain was already known to produce a spirocyclic diketopiperazine alkaloid and cyclotryprostatin B when cultivated on rice medium. More fermentation conditions were experimented and it was eventually discovered that three new and unusual indole-diketopiperazines **1**–**3** were produced by fermentation in a rice medium supplemented with 1% $MgCl_2$ (Figure 1). The proposed BGC is likely responsible for the synthesis of these metabolites. Compound **1** was cytotoxic on HL-60 cells (6.0 μM), and compounds **2** and **3** were active on MCF-7 cell line (7.6 and 10.8 μM, respectively). The marine-derived fungus *Aspergillus fumigatus* CUGBMF170049 was also isolated from the Bohai sea sediments and was shown to produce new pseurotin analogs **4** and **5** (Figure 1) [13]. These metabolites were identified by thorough analysis of spectroscopic data, but unfortunately, were not active in the antimicrobial assays used in this work. Incidentally, the known compound helvolic acid isolated from the same fungus was very active on both *Staphylococcus aureus* and methicillin-resistant *S. aureus*, with a minimal inhibitory concentration (MIC) of 0.78 μg/mL. The fungus *Geosmithia pallida* FS140 was isolated from a sediment collected at 2403 m depth in the South China Sea [14]. Twelve diketopiperazines, including the three new thiodiketopiperazines geospallins A-C (**6**–**8**), were isolated from this fungus (Figure 1). The three new compounds exhibited substantial angiotensin-converting enzyme inhibitory activity, with IC_{50} values of 29–35 µM.

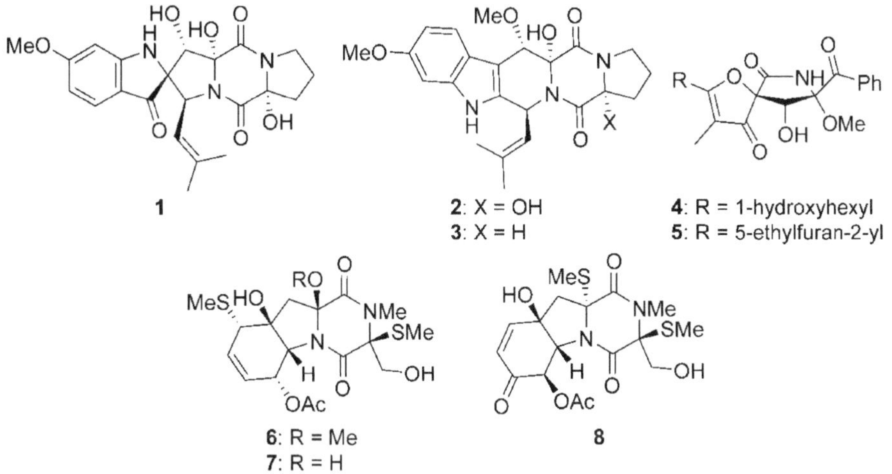

Figure 1. Compounds isolated from marine-sediment-derived fungi.

In the sea, many fungal species are associated with macroscopic organisms [9,11,15]. The isolation of marine fungi from macroorganisms' tissues may in fact be a fairly straightforward way of isolating marine fungi sensu stricto. The fungus *Aspergillus flocculosus* 168ST-16.1 was isolated from the algae *Padina* sp., collected at a depth of 10 m in Da Nang, Vietnam [16]. Five new sesterterpene ophiobolins together with four known ones were isolated from a broth culture of this fungus and identified by spectroscopic methods. Interestingly, all these compounds were active against six cancer cell lines with 50% growth inhibition (GI_{50}) values in the range of 0.14 to 2.01 µM. The new compound 14,15-dehydro-6-*epi*-ophiobolin K (**9**, Figure 2) was among the most active ones, with a GI_{50} of about 0.2 µM on all tested cell lines. The fungal strain *Botryosphaeria ramosa* L29 was isolated as an endophyte of the mangrove plant species. Four new isocoumarin derivatives, namely botryospyrones A, B (**10**), C, and D (Figure 2) were isolated from this fungus. Interestingly, a fifth compound (**11**) was produced only when the culture medium was supplemented with the host plant flavonoid (2R,3R)-3,5,7-trihydroxyflavanone 3-acetate, which had been found to slightly inhibit the endophyte growth. Owing to the ease of isomerization of animals by successive ring-opening and ring-closing steps, it is expected that such compound would eventually generate the more stable stereoisomer, i.e., the *cis* ring junction isomer. Further, the ^{13}C NMR spectrum of compound **11** is similar enough to the one previously reported in the literature to conclude that the proposed structure was probably erroneous [17]. Compound **11** is in fact likely the (3aR,8aS)-1-acetyl-1,2,3,3a,8,8a-hexahydropyrrolo[2,3-*b*]indol-3a-ol rather than the (3aS,8aS) stereoisomer. Three coumarins and compound **11** were evaluated in vitro for antifungal activity toward three phytopathogenic fungi. It is interesting to point out that compound **11**, which is only produced by the fungus exposed to a metabolite of the plant, was the most active, with MIC values in the 28-57 µM range.

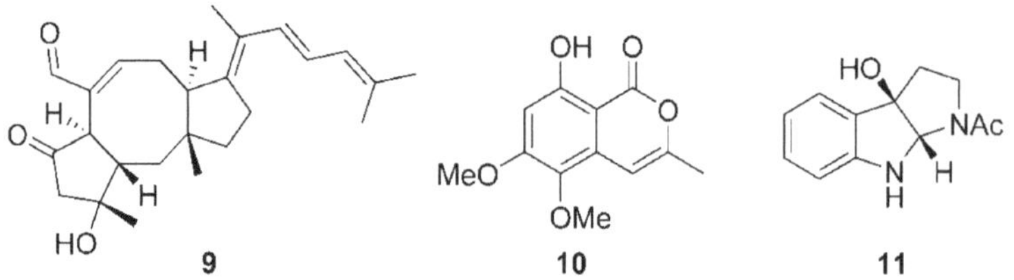

Figure 2. Examples of secondary metabolites isolated from fungi associated to algae and plants.

Finally, describing and understanding the metabolic diversity of marine species will allow a better understanding of the whole marine ecosystem, a better understanding of the mechanisms that regulate interspecific interactions in the sea, and possibly, a better understanding of the evolutionary processes involved. Moreover, the sea is the cauldron of a great diversity of useful and valuable compounds, whether known or not. As such, many marine microorganisms can become important sources of compounds useful to humankind. The carotenoid-producing yeast strain *Rhodotorula* sp. RY1801 was isolated from the exposed intertidal zone along the South Yellow Sea in Dongtai, China [18]. The culture conditions were optimized to increase the microbial biomass and the carotenoid yield of the strain. Eventually, it was possible to reach nearly 1 mg/L carotenoids in the culture, demonstrating that this fast growing *Rhodotorula* sp. strain may be used for the commercial production of carotenoids. The comparison of the phylogenetic and metabolomic profiles of 12 microalgal species from different lineages provided novel insights into the potential of chemotaxonomy in marine phytoplankton [19]. The most abundant and diversified metabolites identified over the 12 strains were polar lipids and pigments. Historically, analysis of pigment composition has been used to assist the classification of microalgae. Here, it was found that lipid classes and some specific lipids within classes may serve as phylogenetic markers. For example, within the tested Mamiellales, the *Ostreococcus* genus differs by the presence of two monogalactosyldiacylglycerols (MGDG 20:5/16:3 and 16:1/16:1), and the species *O. tauri* has six 1,2-diacylglyceryl-3-O-4'-(*N,N,N*-trimethyl)-homoserines (DGTSs) that are not found in *O. mediterraneus*. Overall, a good overlap of phylogenetic and chemotaxonomic signals was demonstrated, in particular when the analysis focused on the major metabolites of algae, and evolutionary divergence between species could be inferred in good congruence with the phylogenies. These results support the hypothesis of a metabolomics equivalent to the "molecular clock" based on the analysis of sequence data.

As seen from the above synopsis, the papers included in this Special Issue provide an interesting overview of chemical diversity of marine fungi and algae. New, bioactive, marine-derived fungal metabolites were isolated and characterized. The articles presented in this Special Issue underline the central role of the sea as a provider of valuable chemicals for human use.

In conclusion, the Guest Editor thanks all the authors who contributed to this Special Issue, all the reviewers for evaluating the submitted manuscripts, and the Editorial board of Marine Drugs, especially, Orazio Taglialatela-Scafati, Editor-in-Chief of the journal, and Estelle Fan, Assistant Editor, for their kind help in bringing this book into reality.

Conflicts of Interest: The author declares no conflict of interest.

References

1. Locey, K.J.; Lennon, J.T. Scaling laws predict global microbial diversity. *Proc. Natl. Acad. Sci. USA* **2016**, *113*, 5970–5975. [CrossRef] [PubMed]
2. Courtois, E.A.; Dexter, K.G.; Paine, C.E.T.; Stien, D.; Engel, J.; Baraloto, C.; Chave, J. Evolutionary patterns of volatile terpene emissions across 202 tropical tree species. *Ecol. Evol.* **2016**, *6*, 2854–2864. [CrossRef] [PubMed]

3. Pye, C.R.; Bertin, M.J.; Lokey, R.S.; Gerwick, W.H.; Linington, R.G. Retrospective analysis of natural products provides insights for future discovery trends. *Proc. Natl. Acad. Sci. USA* **2017**, *114*, 5601–5606. [CrossRef] [PubMed]
4. Wolfender, J.-L.; Litaudon, M.; Touboul, D.; Ferreira Queiroz, E. Innovative omics-based approaches for prioritisation and targeted isolation of natural products – new strategies for drug discovery. *Nat. Prod. Rep.* **2019**, *36*, 855–868. [CrossRef] [PubMed]
5. Gerwick, W.H.; Moore, B.S. Lessons from the past and charting the future of marine natural products drug discovery and chemical biology. *Chem. Biol.* **2012**, *19*, 85–98. [CrossRef] [PubMed]
6. Lauritano, C.; Ferrante, M.I.; Rogato, A. Marine natural products from microalgae: An -omics overview. *Mar. Drugs* **2019**, *17*, 269. [CrossRef] [PubMed]
7. Rodrigues, A.M.S.; Eparvier, V.; Odonne, G.; Amusant, N.; Stien, D.; Houël, E. The antifungal potential of (Z)-ligustilide and the protective effect of eugenol demonstrated by a chemometric approach. *Sci. Rep.* **2019**, *9*, 8729. [CrossRef] [PubMed]
8. Gladfelter, A.S.; James, T.Y.; Amend, A.S. Marine fungi. *Curr. Biol.* **2019**, *29*, R191–R195. [CrossRef] [PubMed]
9. Heo, Y.M.; Kim, K.; Ryu, S.M.; Kwon, S.L.; Park, M.Y.; Kang, J.E.; Hong, J.-H.; Lim, Y.W.; Kim, C.; Kim, B.S.; et al. Diversity and ecology of marine algicolous *Arthrinium* species as a source of bioactive natural products. *Mar. Drugs* **2018**, *16*, 508. [CrossRef]
10. Heo, Y.M.; Lee, H.; Kim, K.; Kwon, S.L.; Park, M.Y.; Kang, J.E.; Kim, G.-H.; Kim, B.S.; Kim, J.-J. Fungal diversity in intertidal mudflats and abandoned solar salterns as a source for biological resources. *Mar. Drugs* **2019**, *17*, 601. [CrossRef]
11. Overy, D.P.; Rämä, T.; Oosterhuis, R.; Walker, A.K.; Pang, K.-L. The neglected marine fungi, *sensu stricto*, and their isolation for natural products' discovery. *Mar. Drugs* **2019**, *17*, 42. [CrossRef] [PubMed]
12. Zhang, Y.-H.; Geng, C.; Zhang, X.-W.; Zhu, H.-J.; Shao, C.-L.; Cao, F.; Wang, C.-Y. Discovery of bioactive indole-diketopiperazines from the marine-derived fungus *Penicillium brasilianum* aided by genomic information. *Mar. Drugs* **2019**, *17*, 514. [CrossRef] [PubMed]
13. Xu, X.; Han, J.; Wang, Y.; Lin, R.; Yang, H.; Li, J.; Wei, S.; Polyak, S.W.; Song, F. Two new spiro-heterocyclic γ-lactams from a marine-derived *Aspergillus fumigatus* strain CUGBMF170049. *Mar. Drugs* **2019**, *17*, 289. [CrossRef] [PubMed]
14. Sun, Z.-H.; Gu, J.; Ye, W.; Wen, L.-X.; Lin, Q.-B.; Li, S.-N.; Chen, Y.-C.; Li, H.-H.; Zhang, W.-M. Geospallins A–C: new thiodiketopiperazines with inhibitory activity against angiotensin-converting enzyme from a deep-sea-derived fungus *Geosmithia pallida* FS140. *Mar. Drugs* **2018**, *16*, 464. [CrossRef] [PubMed]
15. Rédou, V.; Vallet, M.; Meslet-Cladière, L.; Kumar, A.; Pang, K.-L.; Pouchus, Y.-F.; Barbier, G.; Grovel, O.; Bertrand, S.; Prado, S.; et al. Marine fungi. In *The marine microbiome*; Stal, L.J., Cretoiu, M.S., Eds.; Springer International Publishing: Cham, Switzerland, 2016; pp. 99–153. ISBN 978-3-319-33000-6.
16. Choi, B.-K.; Trinh, P.T.H.; Lee, H.-S.; Choi, B.-W.; Kang, J.S.; Ngoc, N.T.D.; Van, T.T.T.; Shin, H.J. New ophiobolin derivatives from the marine fungus *Aspergillus flocculosus* and their cytotoxicities against cancer cells. *Mar. Drugs* **2019**, *17*, 346. [CrossRef] [PubMed]
17. Yang, S.-W.; Cordell, G.A. Metabolism studies of indole derivatives using a staurosporine producer, *Streptomyces staurosporeus*. *J. Nat. Prod.* **1997**, *60*, 44–48. [CrossRef] [PubMed]
18. Zhao, Y.; Guo, L.; Xia, Y.; Zhuang, X.; Chu, W. Isolation, identification of carotenoid-producing *Rhodotorula* sp. from marine environment and optimization for carotenoid production. *Mar. Drugs* **2019**, *17*, 161. [CrossRef] [PubMed]
19. Marcellin-Gros, R.; Piganeau, G.; Stien, D. Metabolomic insights into marine phytoplankton diversity. *Mar. Drugs* **2020**, *18*, 78. [CrossRef] [PubMed]

 *marine drugs*

Review

The Neglected Marine Fungi, *Sensu stricto*, and Their Isolation for Natural Products' Discovery

David P. Overy [1,*], Teppo Rämä [2], Rylee Oosterhuis [3], Allison K. Walker [3] and Ka-Lai Pang [4]

1   Ottawa Research and Development Centre, Agriculture and AgriFood Canada,
    Ottawa, ON K1A 0C6, Canada
2   Marbio, Norwegian College of Fishery Science, University of Tromsø—The Arctic University of Norway,
    9019 Tromsø, Norway; teppo.rama@uit.no
3   Department of Biology, Acadia University, Wolfville, NS B4P2R6, Canada; 117135o@acadiau.ca (R.O.);
    allison.walker@acadiau.ca (A.W.)
4   Institute of Marine Biology and Centre of Excellence for the Oceans, National Taiwan Ocean University,
    20224 Keelung, Taiwan; klpang@mail.ntou.edu.tw
*   Correspondence: david.overy@canada.ca; Tel.: +1-613-759-1857

Received: 11 December 2018; Accepted: 22 December 2018; Published: 10 January 2019

Abstract: Despite the rapid development of molecular techniques relevant for natural product research, culture isolates remain the primary source from which natural products chemists discover and obtain new molecules from microbial sources. Techniques for obtaining and identifying microbial isolates (such as filamentous fungi) are thus of crucial importance for a successful natural products' discovery program. This review is presented as a "best-practices guide" to the collection and isolation of marine fungi for natural products research. Many of these practices are proven techniques used by mycologists for the isolation of a broad diversity of fungi, while others, such as the construction of marine baiting stations and the collection and processing of sea foam using dilution to extinction plating techniques, are methodological adaptations for specialized use in marine/aquatic environments. To this day, marine fungi, *Sensu stricto*, remain one of the few underexplored resources of natural products. Cultivability is one of the main limitations hindering the discovery of natural products from marine fungi. Through encouraged collaboration with marine mycologists and the sharing of historically proven mycological practices for the isolation of marine fungi, our goal is to provide natural products chemists with the necessary tools to explore this resource in-depth and discover new and potentially novel natural products.

Keywords: marine fungi; isolation; culturing; identification; natural products; secondary metabolites

---

1. Introduction

The field of natural product discovery is under transition as genome sequencing and different 'omics' techniques gain more prominence within the research field. Techniques such as transcriptomics and metabolomics, which are used to reveal expressed genes and associated metabolites, are now being applied as tools in the discovery and expression of natural products. 'Omics'-based natural product discovery pipelines are still in their infancy, but have shown their potential in finding new bioactive molecules without culturing the producing organism [1]. However, the isolation and culturing of filamentous fungi and other microorganisms will continue to have relevance in the field for many more years to come and, for many applications, will remain a necessity. For this reason, current and future generations of natural product chemists should know how to isolate, maintain, and preserve fungal cultures.

Filamentous fungi are a proven source of structurally diverse natural products. Few opportunities to explore truly virgin lineages of filamentous fungi for natural products remain; marine fungi in the strict sense (*Sensu stricto, s.s.*) comprise one of these still relatively underexplored areas. Marine fungi *s.s.* are those fungi that are exclusively found to occur in the marine environment, many of which have evolved specialized adaptations that aid in spore dispersal in marine waters. Reports of chemistry from marine fungi *s.s.* suggest that they are an excellent source of new chemical entities, often associated with a variety of different biological activities [2]. For this reason, over the past few decades, many research groups have focused their attention on the marine environment as a source of fungi for natural products discovery. As a result, an exponential increase in the publication of natural products from "marine-derived" fungi has occurred [3,4]. Unfortunately, the majority of most frequently cited marine-derived taxa belong to well-known osmotolerant/halotolerant generalist genera associated with ruderal substratum relationships, such as *Aspergillus* and *Penicillium*, while the discovery rate of new natural products from acknowledged marine fungi *s.s.* has not increased since the late 1990s [2]. Why the disparity?

Cultivability remains one of the main limitations associated with the discovery of natural products from marine fungi *s.s.* [5]. Isolation techniques commonly reported in the literature by marine natural products groups are dominated by simple plating methods, approaches that favor the isolation of faster growing generalist, osmotolerant/halotolerant fungal genera. These methods are antiquated and do not differ significantly from those practiced in the 1940's and 50's [2,6]. The implementation of focused strategies aimed at the isolation of marine fungi *s.s.* are required to tap into this underexplored resource and address the goal of discovering new bioactive natural products. The majority of natural products isolated from marine fungi *s.s.* have resulted from collaborations between natural products chemists and mycologists, where specialized strategies for the isolation of marine fungi *s.s.* were applied. The motivation of this review is to present a guide of best practices for the marine natural products community to implement and increase the discovery rate of natural products from exceptional marine fungi *s.s.* (summarized in Table 1).

Table 1. Summary of isolation techniques of marine fungi *Sensu stricto* presented.

Technique/Substrate	Substrate	References	Notes
Isolation from sea foam	Seafoam	9,11,43–45	Especially suitable for arenicolous species, can yield generalist fungi
Direct plating	Any	41,46	Important to employ multiple media to ensure growth of targeted fungus
Particle filtration and dilution to extinction plating	Any	2,47–50	Especially suited for the isolation of slower growing, unique fungi
Damp chambers	Driftwood, various of macroorganisms	40–42,51–53	Important to employ preventative measures to avoid mite contamination
Baiting stations	Wood	54–55	Encrusting invertebrates can be problematic
In situ culturing	Any	56–60	Novel approach, has the potential to isolate truly unique species

2. Sources of Marine Fungi

Filamentous fungi are major decomposers of woody and herbaceous substrata entering marine ecosystems and propagate by spores, either asexual conidia or by sexual ascospores, basidiospores, or zygospores. In marine ecosystems, spores are dispersed and passively transported through water, ultimately landing on a suitable substrate where they germinate and colonize. During colonization, hyphae (fungal filaments) grow throughout the substrate and, when environmental conditions are appropriate, differentiate into specialized structures to reproduce. Observations of fruiting structures, such as pycnidia or ascomata (typically observed as round "balls" or "tiny dots" emerging from or forming on the substrate surface) and conidiophores ("branch-like" structures bearing wet or dry masses of conidia), are sought after by mycologists for isolating and identifying marine fungi (Figure 1).

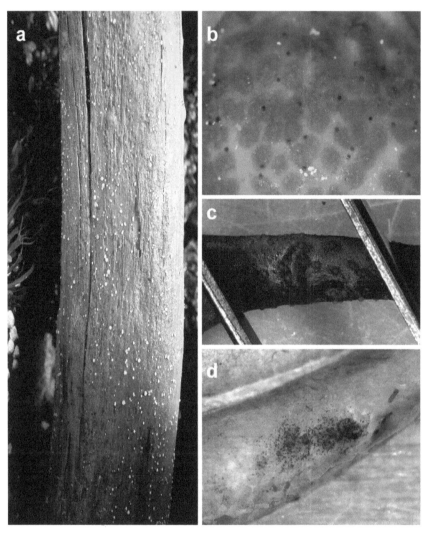

Figure 1. Fruiting structures of marine fungi: (**a**) *Amylocarpus encephaloides* is one of the few species of marine fungi that can be spotted and identified with the naked eye already in the field; (**b**) these "black dots" on the receptacle of macroalga *Ascophyllum nodosum* are fruiting bodies of *Stigmidium ascophylli* (syn. *Mycosphaerella/Mycophycias ascophylli*); (**c**) apothecial fruiting bodies of *Calycina marina* on dead and decaying *A. nodosum*; (**d**) conidial mass of the asexual form of the fungus, *Lulworana uniseptata* (=*Zalerion maritima*), in a borehole made by a *Xylophaga* bivalve species.

Marine fungi are found on a diverse range of substrates: Submerged wood, estuarine plants (such as mangroves and salt marsh grasses and rushes), sand grains, sediment, algae, animals (invertebrates and vertebrates), plankton, and probably even the plastisphere [7,8]. Unlike "marine-derived fungi", many marine fungi *s.s.* have evolved specialized adaptive strategies to ensure successful spore dispersal in the marine environment. Ascomata of arenicolous marine fungi lack long central necks (that would be abraded by the constant movement of the sand grains) and associated ascospores are appendaged and often found to be trapped in sea foam [7,9]. Submerged wood substrata tend to favor the growth of members of the Halosphaeriaceae, Torpedosporales, and Lulworthiales (taxonomic groups in the fungal kingdom composed predominantly of marine fungi), especially in open oceans, where species are typically characterized by passive ascospore release (deliquescing asci) and appendaged ascospores that aid in floatation and attachment [10]. When occurring on submerged wood, ascomata are found partly or completely immersed within the substrata (Figure 2) [9]. Marine Ascomycota are adapted to life in the intertidal zone, and often release spores through the ostiole during periods of low tide, which rest upon the ascomata and become washed off into the water column during high tide, and their ascospores often have gelatinous sheaths (Dothideomycetes), appendages (Sordariomycetes), or, in some species, both to aid in the attachment of spores to substrata [7]. Many conidia (asexual spores) of marine fungi are branched or ornamented or extremely long in size to aid in dispersal and floatation, many of which can be found trapped in seafoam along the shoreline [11].

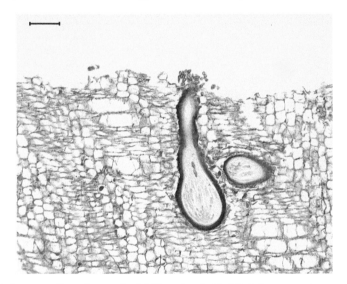

Figure 2. A cross-section micrograph of driftwood-embedded *Dyfrolomyces marinospora* ascomata (fruiting bodies) releasing ascospores (scale bar = 100 μm).

Marine fungi abound in intertidal zones, such as mangrove ecosystems, due to the associated and abundant organic detritus; over 280 species of fungi have been described from submerged mangrove substrata [7]. Algae dominate marine habitats in temperate regions and approximately 80 fungal species have been associated with algae either as parasites or symbionts [12]; however, algae as a source of marine fungal isolates have been grossly neglected and further isolation efforts will likely yield undiscovered species diversity [7]. Estuarine (or salt) marshes occur in the intertidal zone and are dominated by the growth of salt-tolerant plants. Few species of salt-tolerant plants have been surveyed; *Juncus roemerianus* and *Spartina* species are an exception (due to their dominance in this ecosystem) and are associated with a rich diversity of fungal species [13–17].

Although marine fungi in temperate and tropical areas are the most studied, a rich and diverse mycota exist in polar areas [18–22]. Fungi in these cold environments are associated with similar substrate relationships as in warmer regions. Sea-ice, however, is a habitat unique to polar marine regions and supports a diverse community of microorganisms, including fungi. Members of

Chytridiomycota and Dikarya (Ascomycota and Basidiomycota) can be found dwelling in high abundance in sea-ice [18–20]. Studies targeting fungi in macroalgae in Antarctica have revealed high fungal richness [21]. In the Arctic, 100 species of marine fungi have been morphologically detected, whereas molecular data indicates the presence of thousands of fungi in the Arctic marine environment [22].

3. Natural Products from Marine Fungi (*Sensu stricto*)

The phylogenetic breadth of marine fungi *s.s.* within the fungal kingdom is extensive, including marine-specific yeasts, chytrids, and other basal fungal groups; however, fungi in the Dikarya (ascomycetes and basidiomycetes) and the Zoopagomycotina are richer in the number and diversity of secondary metabolite gene clusters and therefore are those most likely to yield the greatest diversity of natural products [23–25]. Filamentous marine fungi (ascomycetes in particular) have proven to be a plentiful source of new natural products (for a detailed review refer to Overy et al., 2014) [2]. For example, from an isolate of *Aigialus parvus* (a marine ascomycete occurring on submerged wood), investigators obtained a total of 11 new natural products representing several different structural classes (aigialomycins A–G, aigialospirol and associated derivatives, and aigialone) [26–31]. *Halorosellinia oceanica* (syn. *Hypoxylon oceanicum*), a marine ascomycete associated with intertidal wood in mangrove habitats, studied by Wyeth for associated antifungal properties, led to the discovery of three new lipodepsipeptides (15G256ε, 15G256γ, 15G256δ), five new macrocyclic polyethers (15G256α, 15G256β, 15G256α-1, 15G256ι, and 15G256ω), and five new linear polyesters (15G256α-2, 15G256β-2, 15G256o, 15G256ν, and 15G256π) [32]. New natural products obtained from marine fungi are associated with a range of biological activities (representative structures are presented in Figure 3). For example, an isolate of the obligate marine species, *Zopfiella marina*, isolated from marine mud (obtained from a depth of 200 m) was found to produce the potent antifungal metabolite, zofimarin, a sordarin derivative [33,34]. Aigialomycin D demonstrated antiplasmodial activity and cytotoxicity against human cell lines (where cytotoxicity has been attributed to cyclin-dependent kinase/glycogen synthase kinase-3 inhibition) [26,35]. The metabolite, pulchellalactam, obtained from a *Corollospora pulchella* strain isolated from driftwood exhibited inhibitory activity against CD45 [36], an essential transmembrane protein tyrosine phosphatase associated with T- and B-cell antigen receptor signaling. Discovery work involving another marine fungus, *Phaeosphaeria spartinae* (=*Leptosphaeria spartinae*), isolated from the inner tissues of the marine alga, *Ceramium* sp., resulted in the characterization of eight new natural products, including spartinoxide and spartinol C, demonstrating inhibition of the enzyme human leukocyte elastase; a disease target associated with pulmonary emphysema, rheumatoid arthritis, and cystic fibrosis [37,38]. These examples emphasize that marine fungi *s.s.* can be isolated from a range of different substrata and can produce unique and previously undiscovered natural products. The following text will provide information regarding several methods that are used by mycologists (and some natural product chemists) to isolate truly unique isolates of marine fungi *s.s.*

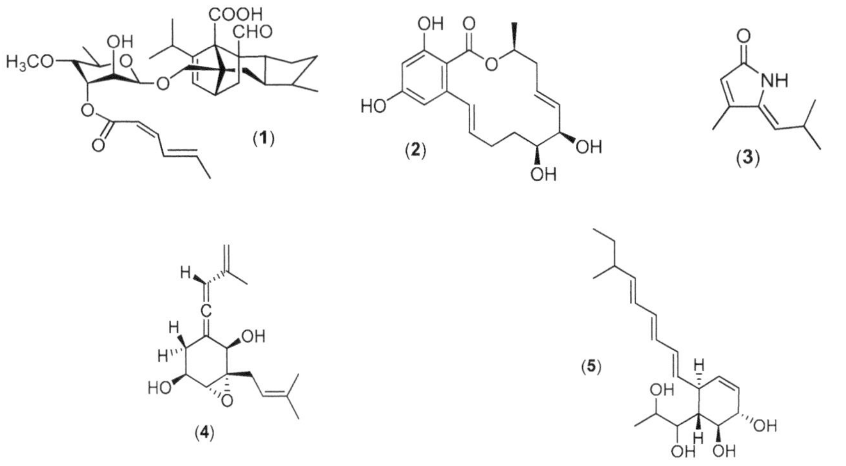

Figure 3. New natural products of marine fungi s.s.: zofimarin (**1**), aigialomycin D (**2**), pulchellalactam (**3**), spartinoxide (**4**) and spartinol C (**5**).

4. Isolation Techniques

4.1. Sample Collection

To obtain marine fungi, marine environments with limited human disturbance are preferable collection sites as they support a greater diversity and denser distribution of fungal species [39]. When obtaining samples, aged materials submerged in seawater with periods of exposure in the air are preferable, e.g., as evidenced by the presence of other marine decay organisms, such as barnacles [9]. For wood and macroalgae, samples with a roughened and softened surface may indicate the presence of fungal growth. On rocky shores, wood trapped between crevices of rocks below the tide line support good fungal growth. For sandy beaches, wood/macroalgae buried deep in the sand or the sand grains occurring near these organic substrata are ideal. In mangrove/salt marsh environments, intertidal substrates (i.e., decaying plant parts either attached to the standing plants or those detached and exposed on the sediment floor during low tide) are often teeming with fruiting bodies of marine fungi. Different stages of decay support different fungal species; therefore, an effort should also be made to collect samples from multiple decay stages [40]. Upon collection, samples should immediately be placed in sterile sealed plastic bags containing paper towel wetted with sterile seawater to avoid desiccation [41]. When collected by SCUBA (self-contained underwater breathing apparatus) diving, samples should be placed in sterile sealable plastic tubes or plastic bags while underwater to minimize handling at the surface (to reduce likelihood of exposure to airborne spores). Samples can then be placed on ice for transport to the laboratory for microscopic examination and culturing. It should be noted that slow drying of samples should be avoided prior to processing as the drying process can cause discharge of fungal ascospores [40]. After returning to the lab, if surface fouling organisms are present on samples, they can be scraped off to prevent bacterial growth and decay. Samples should be washed thoroughly in sterile water to remove thick sediment layers and other potentially contaminating debris. Following rinsing, samples should be drained for approximately one hour to remove any excess water [40]. Samples can then be directly examined under a dissecting microscope or placed in an incubation chamber and examined weekly for the presence of fungal reproductive structures, with weekly spraying with sterile seawater to prevent sample desiccation [42].

Sea foam has been historically used to isolate the highly distinctive appendaged conidia of marine arenicolous (sand-dwelling) ascomycetes [9,43–45]. During sea foam formation, a large number of fungal propagules become occluded within the air bubbles formed as the waves break upon the shoreline and the bubbles percolate through the shoreline sand [45]. A caveat of using sea foam as an isolation substrate is that fungal propagules present in sea foam are not limited to marine fungi in particular; rather, conidia of terrestrial species (saprophytic and plant pathogenic genera) are often blown in from the shoreline and cosmopolitan generalist fungi naturally growing in the intertidal zone are also frequently isolated [11,43]. Sea foam can be collected by passing through a fine mesh sieve (holes approx. 2mm in diam.) using sterile sea water and collected in a sterile container or collected directly into sterile plastic bottles or centrifuge tubes (Figure 4) [9,11,44]. If the foam has collapsed and dried (such as deposition at high tide), it can be carefully scraped off and resuspended in sterile seawater [9]. The addition of an equivalent portion of a double strength preparation of antibiotic solution (in sterile seawater) prior to direct plating or dilution to extinction culturing is advantageous to limit the growth of bacteria [11].

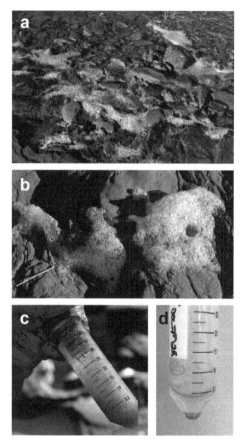

Figure 4. Seafoam as a source of inoculum: (**a,b**) Seafoam accumulating onshore during low tide; (**c**) collection of seafoam into sterile centrifuge tubes (care should be taken not to collect any additional sediment or water); (**d**) upon return to lab, a double strength antibiotic in sterile seawater should be added and the seafoam allowed to settle for several hours at 4 °C. Particles/spores can then be concentrated by centrifuging for dilution plating. Note: Storage of samples for longer than a day prior to processing is not recommended, as marine yeasts will continue to grow in the solution, even at refrigeration temperatures and will hamper the isolation of axenic cultures of filamentous fungi.

4.2. Direct Plating vs. Particle Filtration and Dilution to Extinction Plating

Direct plating techniques offer a means to capture marine fungi that are embedded in their respective substrata and may not be apparent upon visual inspection in the lab. This method is extremely useful for large substrata, such as submerged wood samples, prop roots of mangrove plants, algae, and the stems/roots of salt marsh plants. Fragments (approximately 0.5 cm^2) can be aseptically cut and placed into petri dishes containing sterile seawater and antibiotics for two to three hours. These fragments can then be subsequently sliced further (to a size of approximately 3 mm^2) and transferred to agar plates overlaid with sterile seawater and antibiotics for incubation [41].

Alternatively, samples can be collected in the field (which is ideal when sampling from a substrate that is large and not easily transported to the lab, such as large driftwood samples or logs [46]). In the field, thin slices of wood are cut-off from the surface wood using EtOH and flame sterilized knife and forceps (Figure 5). From the spot where the surface wood is removed, a wooden cube is cut-off and transferred with forceps into sterile plastic bags that are sealed airtight. The sample is kept cool, transported to the laboratory, and plated on agar plates. The plates are monitored over a period of up to several months to allow the appearance of the slowest growing fungi. Mycelia growing out of the wooden cubes are transferred to fresh agar plates and sub-cultured until axenic cultures are obtained. The incubation conditions should mimic the natural conditions as much as possible to select for marine fungi *s.s.* This method has been shown to result in an approximately half of pure culture isolates being marine fungi, including a significant portion of *Sensu stricto* species [46].

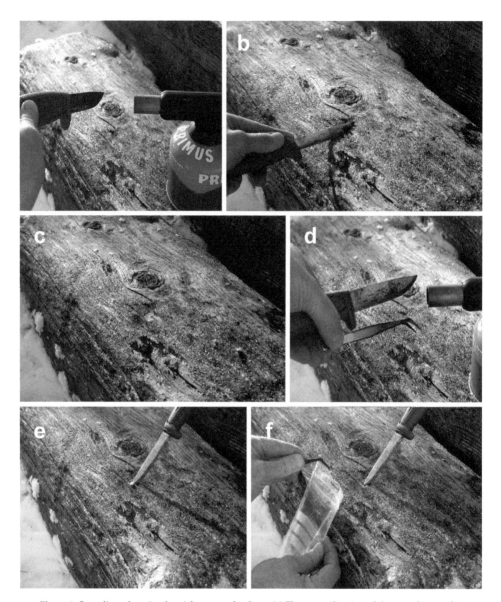

Figure 5. Sampling of marine fungi from wooden logs: (**a**) Flame-sterilization of the sampling knife sprayed with EtOH using a gas burner; (**b**) cutting off a thin slice of surface wood to exclude surface contaminants (using a sharp knife, take as thin a slice as possible to avoid excluding marine fungi inhabiting the upper cell layers below the wood surface); (**c**) a sterile spot on the wooden log prepared for sampling; (**d**) surface sterilization of the knife and forceps with EtOH and the gas burner; (**e**) cutting of a small piece of wood from the wood; (**f**) picking up the wooden piece with forceps and placing it into a sterile plastic bag that is sealed airtight.

The combination of substrata particle filtration and dilution-to extinction particle plating is an improvement upon more traditional isolation approaches and has been proven as being effective in obtaining marine fungi and thus merits more wide-spread adoption (Figure 6) [2,47]. Substrata deconstructed by homogenization are filtered and washed through a series of meshes of decreasing pore sizes and particles of a given size are harvested and dispensed at high-dilution rates into 48- or 96-well plates [48–50]. This technique increases the rate of capture of fungi originating from vegetative fragments embedded in the substrata; more importantly, partitioning into 48- or 96- well plates allows for slower growing fungi (such as obligate marine fungi) actively growing within substrata particles to be separated and isolated from faster growing taxa. The particle filtration/dilution to extinction culturing approach is ideal for working with marine sediments. Dilution culturing into 48-well plates is a proven method for the successful isolation of marine fungi *s.s.* from seafoam suspensions [11].

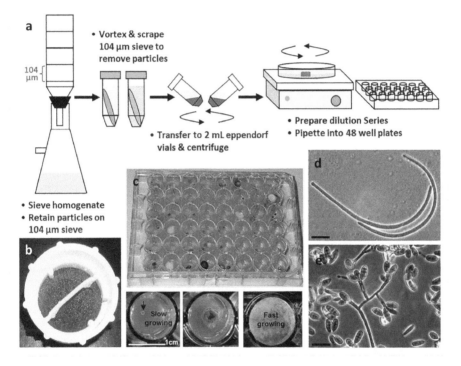

Figure 6. The use of particle filtration and dilution culturing in 48-well plates for the isolation of marine fungi (Overy et al. 2014 [2]): (**a**) Schematic of the particle filtration and dilution plating workflow; (**b**) algal particles that were retained on a filter sieve after homogenization and filtration; (**c**) plating into 48-well plates allows for the separation of fast and slower growing isolates, increasing the isolation frequency of slower growing fungi; (**d**) conidia from a strain of marine fungus, *Anguillospora marina* (telemorph = *Lindra obtusa*), isolated from algal tissue particles (scale bar = 20 μm); (**e**) conidia and conidiophores of a strain of the marine fungus, *Paradendryphiella arenaria*, isolated from algal tissue particles (scale bar = 20 μm).

4.3. Damp Chambers

The incubation of samples in damp chambers provide a moist (high water activity) environment to encourage fungal growth and the formation of sexual fruiting structures and conidiophores in a laboratory setting. Damp chambers consist of a sterilized plastic container (i.e., a Tupperware box, Petri dish, or plastic bag) lined with a sterile, absorbent material (i.e., filter paper, sand, or vermiculite), which is moistened with sterile seawater (Figure 7). Sand or vermiculite are preferable as they are poor growth substrates for bacteria. To prevent bacterial growth, damp chambers should be kept free of accumulated water (and if filter paper is used, it should be moistened with sterile seawater containing antibiotics) [41]. It is important to note that the damp chambering of environmental samples can lead to problematic infestations from mites and other insects (resulting in cross contamination within and between samples), but can easily be prevented by including a small container of camphor in the damp chamber [40]. Damp chambers should be incubated within 5 °C of the ambient temperature at their collection site and sprayed twice weekly with sterile seawater to prevent desiccation [42]. Recommended incubation periods vary between 6–18 months depending on the sample substratum and its length of decay [51]. Samples should be examined periodically under a dissecting microscope for mycelial outgrowth and the development of fungal fruiting bodies or sporulating conidiophores. For example, wood should be examined frequently for 3 months after collection while leaves must be examined at regular intervals within 2 weeks of collection [41,52]. Due to fungal succession (the sequence of fungi sporulating on the same substrate over time), it is important to examine a specimen frequently to maximize the number of species isolated [53].

Figure 7. Example of a driftwood damp chamber; note the use of a camphor ball for the prevention of mites.

4.4. Baiting Stations

Stations can be created using various baits, such as stems of salt marsh plants or wood (i.e. beech, maple, pine, or balsa wood). Baits should first be sterilized by autoclaving and then strung on a line (i.e., nylon rope or fishing line), a few inches apart, and submerged approximately 0.5 m from the bottom. The use of floats and an appropriate anchor will ensure that the bait lines remain vertical in the water column, do not move their location with the tide, and provide a visual guide to identify their location (Figure 8). The length of submersion time will affect the fungal biodiversity that can be isolated (fungal succession). Over time, baits can also become encrusted with invertebrates, such as barnacles and encrusting tunicates, or become consumed by shipworm, which will hinder the isolation of fungi. One advantage of this technique is that the selection of substrata can be controlled, as hard wood and soft wood can lead to the isolation different species of marine fungi. To harvest, baits are removed from the line and placed into a sterile bag or suitable sterile container for transportation to the lab. In a sterile environment, the bait surface should be scraped with a sterile spatula to remove surface debris and invertebrates (as these organisms will rot and promote significant bacterial growth in the damp chamber) and rinsed several times in sterile seawater. Overnight incubation in plastic bags containing an antibiotic preparation in sterile seawater is effective in reducing the number of viable bacterial cells prior to placing the bait into damp chambers. If using sterile filter paper in the damp chamber, the filter paper should also be saturated with seawater containing antibiotics to prevent the growth of bacteria. Baits should be examined at regular intervals and provided the damp chambers remain humid, be kept for up to a year to monitor for the emergence of different marine fungi [54]. Ideally, damp chambers should be incubated at temperatures proximal to the average temperature range of sea water in which the baiting station was placed [55].

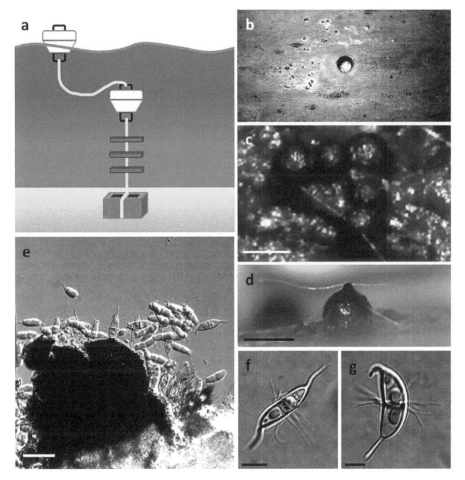

Figure 8. The use of baiting stations for the isolation of marine fungi (Overy et al., 2014 [2]): (**a**) Sterilized wooden boards are suspended in the water column; (**b**) harvested wooden board after 6 months in a damp chamber, ascomata are apparent on the surface; (**c,d**) ascomata of *Corollospora maritima* in situ (scale bar = 250 μm); (**e**) squash mount of *C. maritima* ascomata obtained from board (scale bar = 30 μm); (**f,g**) ascospores of *C. maritima* with characteristic polar and equatorial appendages (scale bar = 10 μm).

5. In Situ Culturing

Direct culturing vs culture independent biodiversity studies predict that only approximately 1% of prokaryotes can be cultured using conventional techniques [56]. Although there are no total estimates for the cultivability of marine fungi, from similar culture vs. culture independent data sets, a low culture recovery (5–7%) has also been observed (Rämä unpublished). Culturing conditions in the laboratory are gross simplifications of natural systems. Many microbes require environmental interactions to grow. Therefore, to access a more bio-diverse range of marine fungi, new and innovative isolation methodologies need to take these factors into consideration. In situ culturing techniques attempt to address these issues and, for bacteriology, have resulted in the successful isolation of previously undescribed organisms [57,58]. Here, suspensions of environmental samples are held between semipermeable membranes and placed back into their native environment. The semipermeable membranes permit the diffusion of growth factors to encourage microbial growth while retaining/preventing the movement or exchange of microbial cells to the environment. It has been shown that microbial cells maintained in in situ culturing devices can, to a significant extent, be domesticated and subsequently subcultured in vitro [57]. There are several existing applications for isolating and cultivating previously uncultivated microorganisms, such as diffusion chambers, its high-throughput version, the Ichip, and the commercially available MicroDish Biochamber [57].

Whereas the previous applications are based on inoculating cells inside each device that will then be taken out to culture in situ, microbial traps operate according to another principle [59]. An empty device, with growth medium sandwiched between membranes allowing movement of growth factors and/or cells, is taken out to a desired environment. During incubation, filamentous organisms grow into the device through the membrane pores and can later be domesticated in pure cultures in vitro. Using a membrane with a pore size of 0.4–0.6 μm will select for fungi and other filamentous organisms [60]. The selection of a small enough pore size blocks all cell movement and can be used to protect the growth medium inside the device against, e.g., airborne contaminants when trapping sediment fungi. Microbial traps are not available commercially, but they are simple in structure and can be easily built (Figure 9). The construction should take place in a sterile environment to avoid getting contaminants inside the trap.

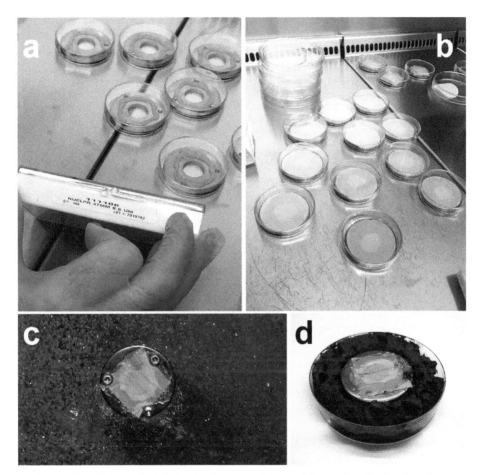

Figure 9. Microbial traps for isolating fungi constructed from stainless steel washer (e.g., 50 × 3 mm in dimensions and with a through whole of 17 mm in diameter) and polycarbonate membranes (e.g., Whatman™ Nuclepore™): (**a**) Steel washers with three drill holes (for securing to solid surfaces, such as driftwood or sea ice), each ready for the construction of a microbial trap under laminar flow; (**b**) using an appropriate adhesive (e.g., aquarium silicon), bottom membranes of the traps with a sufficient pore size to allow fungal filaments to grow into the trap are glued in place (next steps (not pictured) include filling the inside of the washer with agar and, after cooling down of the agar, gluing the bottom membrane of the trap in place); (**c**) using screws, the microbial trap is secured in situ on sea-ice (the figure displays the top of the trap with a small-pored membrane that does not allow movement of the cells from surrounding air into the trap containing agar medium); (**d**) trapping deep-sea sediment fungi using in vitro incubation on sediment sample (again, the top membrane does not allow contaminants to enter the agar medium inside the trap - for a comprehensive description of the construction, see Epstein et al. (2010) [60]).

6. Isolation and Examination of Samples

Fruiting bodies of marine ascomycetes occur superficially, partially, or fully immersed on/in their respective substrates and when viewed under a stereo microscope, can be isolated directly from substrata. The contents of fruiting bodies or conidia/conidiophores can be aseptically transferred using sterilized fine-tipped forceps or a pin to an appropriate culture medium. Fruiting bodies that are completely immersed within a substratum can be exposed using a sterile razor blade or scalpel. Once exposed, the fruiting body centrum can be transferred to an isolation medium or slide for microscopic examination. For single-spore isolations, centrum material should be first aseptically dispersed in drops of sterile seawater on a sterile glass microscope slide. This spore suspension is then transferred to an appropriate culture medium (i.e., seawater or cornmeal agar containing antibiotics) and monitored daily for the germination of spores using a stereomicroscope. Once observed, individual germinated spores can be transferred by a flame sterilized needle onto a new culture medium and repeatedly sub-cultured to produce axenic cultures [40]. The best time to subculture is while colonies are distinct and autonomous, from areas of newest growth. Mycelial colonies from single spore isolations are preferred to ensure reproducibility in liquid fermentation or solid substrate culturing when screening natural products. For asexual marine fungi with conidia observed from conidiophores appearing on the surface of substrates, individual spores can be picked and inoculated directly onto cornmeal agar plates with antibiotics; alternatively, the single spore isolation method described above can be applied. The few described marine Basidiomycota generally produce reduced size basidiomata, which can be crushed to release spores for identification and isolated as described above.

Various media formulations can be employed for the isolation of marine fungi *s.s.*; however, the choice of isolation medium will have a direct impact on the diversity of fungi obtained [61]. Media that are rich in simple sugars may restrict the growth and sporulation of lignicolous fungi; rather, such rich media often promote the growth of more generalist, cosmopolitan genera, such as *Pencillium* and *Aspergillus*. Lignicolous marine fungi prefer a minimal medium, such as 0.1% glucose, 0.01% yeast extract, 1.8% agar in aged seawater, or adding sterilized birch or balsa wood on top of sea water agar [62]. Seawater or cornmeal seawater agar are commonly used alternatives for the isolation of wood-inhabiting fungi as most species germinate on these media. Agar containing mono- or disaccharides should be avoided if yeast growth is a problem. Specialized carbon sources, such as cornmeal seawater agar or V8 seawater agar, can be used to prevent yeast growth [42,52,63]. Including antibiotics to the isolation media prevents bacterial growth; the addition of cyclohexamide will retard faster, growing generalist, cosmopolitan fungi. Penicillin G (300 mg/L) and streptomycin sulfate (500 mg/L) are effective when used together, added to cooled agar immediately before pouring plates; whereas chloramphenicol can be added to the medium prior to autoclaving. Ideally, inoculated agar plates should be incubated at temperatures close to that of the collection site until sporulation is observed [41]. Once stable cultures are obtained, they can then be transferred to more nutrient rich or specialized media designed for natural product production.

Microscopic examination is an essential step in identifying or confirming the taxonomy of the isolated fungus and is strongly encouraged at the point of initial inoculation, as many fungi produce sterile colonies in culture. Fruiting bodies and conidiophores are placed on a glass microscope slide in a drop of appropriate mounting solution (i.e., lactic acid, lactophenol, cotton blue, sterile sea water, etc.) and covered with a glass cover slip for viewing under a compound microscope. In the case of fruiting structures, such as pycnidia or ascomata, gently pressing on the cover slip will cause the fruiting body to rupture and release its spores, creating a 'squash-mount' [9]. As well, centrum contents can be removed directly to a mounting medium for viewing under the microscope. Compound microscopes with a magnification of at least 400 × (ideally 600 × 1000 ×) are best suited for observation of micromorphology as many spores (ie., conidia, ascospores, basidiospores, zygospores, etc.) are less than 20 μm in diameter or length, and information, such as ornamentation, are critical identification characteristics and cannot be resolved at lower magnifications. When using transmitted light microscopy, the use of stains, such as lactophenol cotton blue, aqueous nigrosine, and India ink, can improve contrast. Phase contrast and differential interference contrast microscopy are also useful optical techniques for increasing contrast when viewing specimens. Ascospore appendages and sheaths are often diagnostic for particular species and will only unwind after a period of submersion in water [41]. Dichotomous and pictorial keys based on these fungal reproductive structures are useful identification resources [9,64–66]. Semi-permanent microscope slide mounts can be made following the method of Volkmann-Kohlmeyer and Kohlmeyer [67].

In addition to micromorphology, identification by DNA sequencing is encouraged. The internal transcribed spacer region of ribosomal DNA is the acknowledged species-level DNA barcode region for fungi and should be sequenced and compared with the reference DNA database, NCBI GenBank, using the BLAST tool [68]. For certain groups of fungi, additional genes may be informative for species-level identifications, e.g., β-tubulin or elongation factor 1-α; for marine yeasts, the 28S ribosomal large subunit region; and for basal fungal lineages, the 18S ribosomal subunit region [69]. Obtained DNA sequences should be deposited in GenBank and associated with vouchered specimen material (dried specimen and/or preserved cultures), as our fungal identifications are only as precise as our reference databases [46].

7. Preservation

Cryopreservation is preferred for long-term storage of fungal cultures. Agar discs cut from living cultures are placed into sterilized cryoplastic microtubes containing 1 mL of sterile 10% *w/v* glycerol as a cryoprotectant [62]. Four to five agar discs should be transferred in a laminar flow cabinet from the growing margin of a fungal culture using a sterilized transfer stick or needle. Microtubes should be stored at 15 °C for 15 min before freezing them at a rate of 1 cm per minute from 15 to −40 °C and 2 cm from −40 to −80 °C. Tubes can then be moved to a storage vessel and preserved in liquid nitrogen at −180 °C. Alternative cryoprotectants include 10% dimethylsulfoxide (DMSO) or 10% glycerol 5% trehalose, which vary in effectiveness depending on the fungal species. For shorter-term storage, agar discs obtained from cultures as above can be stored in sterile distilled water at 4 °C with annual replacement.

Subcultures can also be maintained for storage on agar slants. Slants are made in test tubes with 0.1% agar overlaid with sterile seawater or mineral oil [40,62]. Spores or small colonies are transferred to the surface of the slant using aseptic techniques. The test tube is then capped and stored at 15–20 °C for up to 8 weeks [40]. Alternatively, subcultures can be transferred from agar plates onto potato dextrose or corn meal agar test tube slants and stored at 4 °C permanently (with annual replacement). Cultures can also be preserved using lyophilization (freeze-drying). This method of preservation, however, is only suitable for fungi that sporulate abundantly. As a result, it is not ideal for most marine fungi [62]. To avoid loss of activity through continuous subculturing, multiple first-generation cryostocks should be prepared.

8. Concluding Remarks

It can be argued that "marine-derived" fungi are a proven source of new natural products; this is not in dispute as over a thousand new compounds or new derivatives of known molecules have been discovered over the past few decades. However, recent evidence has demonstrated that the new chemistry is a result of saline stress, where terrestrial counterparts of the same species produce the same chemistry as discovered from marine-derived strains when grown on media containing seawater [70]. Simply mining the terrestrial isolates of osmotolerant species on media containing seawater will achieve the same ends. However, in such a scenario, marine fungi (*Sensu stricto*) will remain an underexplored resource. To address this, attention currently focused on marine derived fungi must shift focus to the isolation of marine fungi (*Sensu stricto*). In addition, new culturing techniques should be adopted and developed for accessing a greater diversity of marine fungi that has been shown to exist using environmental sequencing.

Assessment of whether marine fungi have evolved to produce secondary metabolites that are distinct from their terrestrial relatives is awaiting completion of genomic sequences. These data will lead to hypotheses about how isolation in the marine habitat may have influenced the evolution of secondary metabolite gene clusters in terms of divergence from ancestral pathways or loss or expansion of pathways for individual structural classes. The principal ascomycete lineages of marine fungi derive primarily from the Sordariomycetidae, Dothideomycetidae, and the Eurotiomycetidae [71–75]. Sequencing of several hundred fungal genomes of representative terrestrial species revealed that the main subclasses of these subphyla often have very rich secondary metabolomes, therefore, it is logical to assume that the marine lineages derived from within these ascomycete subclasses would also have an active and complex secondary metabolism. The few reports of chemistry from acknowledged marine fungi suggest that they will be an excellent source of new chemical entities, often associated with a variety of different biological activities [2].

There are several examples in the literature where collaborations between marine mycologists and natural product chemists have resulted in new chemistry discovered from marine fungi [27,32,76–79]. Hence, we encourage chemists to collaborate with marine mycologists, many of which have personal collections of marine fungi that can be provided for chemical investigation or to acquire marine fungi from recognized culture collections. Through active cooperation and adoption of classical isolation methodologies with modern adaptations, the field of natural products discovery from marine fungi (*Sensu stricto*) will facilitate a deeper exploration of this truly unique and underexplored group of fungi.

Author Contributions: All authors contributed to the writing—original draft preparation; D.P.O., T.R., A.W. and K.L.P. contributed to the review and editing of the manuscript.

Funding: D.P.O. would like to acknowledge financial support from AAFC project J-001848 and A.W. would like to acknowledge NSERC Discovery Grant 2017-04325.

Acknowledgments: Jeanette H. Andersen is acknowledged for pictures.

Conflicts of Interest: The authors declare no conflict of interest.

References

1. Hover, B.M.; Kim, S.-H.; Katz, M.; Charlop-Powers, Z.; Owen, J.G.; Ternei, M.A.; Maniko, J.; Estrela, A.B.; Molina, H.; Park, S. Culture-independent discovery of the malacidins as calcium-dependent antibiotics with activity against multidrug-resistant Gram-positive pathogen. *Nat. Microbiol.* **2018**, *3*, 415–422. [CrossRef] [PubMed]
2. Overy, D.P.; Bayman, P.; Kerr, R.G.; Bills, G.F. An assessment of natural product discovery from marine (*sensu strictu*) and marine-derived fungi. *Mycology* **2014**, *5*, 145–167. [CrossRef]
3. Bugni, T.S.; Ireland, C.M. Marine-derived fungi: A chemically and biologically diverse group of microorganisms. *Nat. Prod. Rep.* **2004**, *21*, 143–163. [CrossRef] [PubMed]
4. Rateb, M.E.; Ebel, R. Secondary metabolites of fungi from marine habitats. *Nat. Prod. Rep.* **2011**, *28*, 290–344. [CrossRef] [PubMed]
5. Tasdemir, D. Marine fungi in the spotlight: Opportunities and challenges for marine fungal natural product discovery and biotechnology. *Fungal Biol. Biotechnol.* **2017**, *4*, 5. [CrossRef]
6. Karwehl, S.; Stadler, M. Exploitation of fungal biodiversity for discovery of novel antibiotics. In *How to Overcome the Antibiotic Crisis*; Stadler, M., Dersch, P., Eds.; Springer International Publishing AG: Cham, Switzerland, 2016; pp. 303–338.
7. Jones, E.B.G. Fifty years of marine mycology. *Fungal Divers.* **2011**, *50*, 73–112. [CrossRef]
8. Zettler, E.R.; Mincer, T.J.; Amaral-Zettler, L.A. Life in the "Plastisphere": Microbial communities on plastic marine debris. *Environ. Sci. Technol.* **2013**, *47*, 7137–7146. [CrossRef]
9. Kohlmeyer, J.; Kohlmeyer, E. *Marine Mycology: The Higher Fungi*; Academic Press: New York, NY, USA, 1979.
10. Jones, E.B.G. Ultrastructure and taxonomy of the aquatic ascomycetous order *Halosphaeriales*. *Can. J. Bot.* **1995**, *73*, 790–801. [CrossRef]
11. Overy, D.P.; Berrue, F.; Correa, H.; Hanif, N.; Hay, K.; Lanteigne, M.; Mquilian, K.; Duffy, S.; Boland, P.; Jagannathan, R.; et al. Sea foam as a source of fungal inoculum for the isolation of biologically active natural products. *Mycology* **2014**, *5*, 130–144. [CrossRef]
12. Zuccaro, A.; Mitchell, J. Fungal communities of seaweeds. In *The fungal Community: Its Organization and Role in the Ecosystem*; Deighton, J., White, J., Oudemans, P., Eds.; CRC Press: New York, NY, USA, 2005; pp. 553–580.

13. Gessner, R.V.; Goos, R. Fungi from *Spartina alterniflora* in sRhode Island. *Mycologia* **1973**, *65*, 1296–1301. [CrossRef]

14. Newell, S.Y.; Porter, D.; Lingle, W.L. Lignocellulolysis by ascomycetes (fungi) of a saltmarsh grass (smooth cordgrass). *Microsc. Res. Tech.* **1996**, *33*, 32–46. [CrossRef]

15. Kohlmeyer, J.; Volkmann-Kohlmeyer, B. The biodiversity of fungi on *Juncus roemerianus*. *Mycol. Res.* **2001**, *105*, 1411–1412. [CrossRef]

16. Walker, A.K.; Campbell, J. Marine fungal diversity: A comparison of natural and created salt marshes of the north-central Gulf of Mexico. *Mycologia* **2010**, *102*, 513–521. [CrossRef] [PubMed]

17. Elmer, W.H.; Marra, R.E. New species of *Fusarium* associated with dieback of *Spartina alterniflora* in Atlantic salt marshes. *Mycologia* **2011**, *103*, 806–819. [CrossRef] [PubMed]

18. Zhang, T.; Wang, N.-F.; Zhang, Y.-Q.; Liu, H.-Y.; Yu, L.-Y. Diversity and distribution of aquatic fungal communities in the Ny-Ålesund region, Svalbard (high arctic). *Microb. Ecol.* **2016**, *71*, 543–554. [CrossRef] [PubMed]

19. Hassett, B.T.; Ducluzeau, A.-L.L.; Collins, R.E.; Gradinger, R. Spatial distribution of aquatic marine fungi across the western Arctic and sub-arctic. *Environ. Microbiol.* **2017**, *19*, 475–484. [CrossRef]

20. Hassett, B.T.; Gradinger, R. Chytrids dominate arctic marine fungal communities. *Environ. Microbiol.* **2016**, *18*, 2001–2009. [CrossRef]

21. Loque, C.P.; Medeiros, A.O.; Pellizzari, F.M.; Oliveira, E.C.; Rosa, C.A.; Rosa, L.H. Fungal community associated with marine macroalgae from Antarctica. *Polar Biol.* **2010**, *33*, 641–648. [CrossRef]

22. Rämä, T.; Hassett, B.T.; Bubnova, E. Arctic marine fungi from filaments and flagella to operational taxonomic units and beyond. *Bot. Mar.* **2017**, *60*, 433–452. [CrossRef]

23. Arvas, M.; Kivioja, T.; Mitchell, A.; Saloheimo, M.; Ussery, D.; Penttila, M.; Oliver, S. Comparison of protein coding gene contents of the fungal phyla Pezizomycotina and Saccharomycotina. *BMC Genom.* **2007**, *8*, 325. [CrossRef]

24. Lackner, G.; Misiek, M.; Braesel, J.; Hoffmeister, D. Genome mining reveals the evolutionary origin and biosynthetic potential of basidiomycete polyketide synthases. *Fungal Genet. Biol.* **2012**, *49*, 996–1003. [CrossRef] [PubMed]

25. Ohm, R.A.; Feau, N.; Henrissat, B.; Schoch, C.L.; Horwitz, B.A.; Barry, K.W.; Condon, B.J.; Copeland, A.C.; Dhillon, B.; Glaser, F.; et al. Diverse lifestyles and strategies of plant pathogenesis encoded in the genomes of eighteen Dothideomycetes fungi. *PLoS Pathog.* **2012**, *8*, e1003037. [CrossRef] [PubMed]

26. Isaka, M.; Suyarnsestakorn, C.; Tanticharoen, M.; Kongsaeree, P.; Thebtaranonth, Y. Aigialomycins A-E, new resorcylic macrolides from the marine mangrove fungus *Aigialus parvus*. *J. Org. Chem.* **2002**, *67*, 1561–1566. [CrossRef] [PubMed]

27. Isaka, M.; Yangchum, A.; Intamas, S.; Kocharin, K.; Jones, E.G.; Kongsaeree, P.; Prabpai, S. Aigialomycins and related polyketide metabolites from the mangrove fungus *Aigialus parvus* BCC 5311. *Tetrahedron* **2009**, *65*, 4396–4403. [CrossRef]

28. Vongvilai, P.; Isaka, M.; Kittakoop, P.; Srikitikulchai, P.; Kongsaeree, P.; Thebtaranonth, Y. Ketene acetal and spiroacetal constituents of the marine fungus *Aigialus parvus* BCC 5311. *J. Nat. Prod.* **2004**, *67*, 457–460. [CrossRef] [PubMed]

29. Kohlmeyer, J.; Schatz, S. *Aigialus* gen. novo (Ascomycetes) with two marine species from mangroves. *Trans. Br. Mycol. Soc.* **1985**, *85*, 699–707. [CrossRef]

30. Tan, T.K.; Teng, C.L.; Jones, E.B.G. Substrate type and microbial interactions as factors affecting ascocarp formation by mangrove fungi. *Hydrobiologia* **1995**, *295*, 127–134. [CrossRef]

31. Alias, S.; Kuthubutheen, A.; Jones, E.G. Frequency of occurrence of fungi on wood in Malaysian mangroves. In *Asia-Pacific Symposium on Mangrove Ecosystems, Developments in Hydrobiology*; Alias, S.A., Kuthubutheen, A.J., Jones, E.B.G., Eds.; Springer: Dordrecht, The Netherlands, 1995; Volume 106, pp. 97–106.

32. Abbanat, D.; Leighton, M.; Maiese, W.; Jones, E.; Pearce, C.; Greenstein, M. Cell wall active antifungal compounds produced by the marine fungus *Hypoxylon oceanicum* LL-15G256. *J. Antibiot.* **1998**, *51*, 296–302. [CrossRef]

33. Ogita, J.; Hayashi, A.; Sato, S.; Furutani, W. Antibiotic Zopfimarin. Japan Patent 62-40292, 1987.

34. Kondo, M.; Takayama, T.; Furuya, K.; Okudaira, M.; Hayashi, T.; Kinoshita, M. A nuclear magnetic resonance study of Zopfinol isolated from *Zopfiella marina*. *Annu. Rep. Sankyo Res. Lab.* **1987**, *39*, 45–53.

35. Barluenga, S.; Dakas, P.Y.; Ferandin, Y.; Meijer, L.; Winssinger, N. Modular asymmetric synthesis of aigialomycin D, a kinase-inhibitory scaffold. *Angew. Chem. Int. Ed.* **2006**, *45*, 3951–3954. [CrossRef]

36. Alvi, K.A.; Casey, A.; Nair, B.G. Pulchellalactam: A CD45 protein tyrosine phosphatase inhibitor from the marine fungus *Corollospora pulchella*. *J. Antibiot.* **1998**, *51*, 515–517. [CrossRef]

37. Elsebai, M.F.; Kehraus, S.; Gütschow, M.; Koenig, G.M. New polyketides from the marine-derived fungus *Phaeosphaeria spartinae*. *Nat. Prod. Commun.* **2009**, *4*, 1463–1468. [PubMed]

38. Elsebai, M.F.; Kehraus, S.; Gütschow, M.; Koenig, G.M. Spartinoxide, a new enantiomer of A82775C with inhibitory activity toward HLE from the marine-derived fungus *Phaeosphaeria spartinae*. *Nat. Prod. Commun.* **2010**, *5*, 1071–1076. [PubMed]

39. Hyde, K.D.; Jones, E.B.G.; Leaño, E.; Pointing, S.B.; Poonyth, A.D.; Vrijmoed, L.L.P. Role of fungi in marine ecosystems. *Biodivers. Conserv.* **1998**, *7*, 1147–1161. [CrossRef]

40. Bremer, G. Isolation and culture of thraustochytrids. In *Marine Mycology—A Practical Approach*; Hyde, K.D., Pointing, S.B., Eds.; Fungal Diversity Press: Hong Kong, China, 2000; pp. 49–61.

41. Vrijmoed, L.L.P. Isolation and culture of higher filamentous fungi. In *Marine Mycology—A Practical Approach*; Hyde, K.D., Pointing, S.B., Eds.; Fungal Diversity Press: Hong Kong, China, 2000; pp. 1–20.

42. Walker, A.K. Marine Fungi of U.S. Gulf of Mexico Barrier Island Beaches: Biodiversity and Sampling Strategy. Ph.D. Thesis, University of Southern Missisippi, Hattiesburg, MS, USA, December 2012.

43. Koehn, R.D. Fungi isolated from sea foam collected at North Padre island beaches. *Southwest. Nat.* **1982**, *27*, 17–21. [CrossRef]

44. Kirk, P.W. Direct enumeration of marine arenicolous fungi. *Mycologia* **1983**, *75*, 670–682. [CrossRef]

45. Hyde, K.D.; Jones, E.G. Introduction to fungal succession. *Bot. J. Linn. Soc.* **1989**, *100*, 237–254. [CrossRef]

46. Rämä, T.; Nordén, J.; Davey, M.L.; Mathiassen, G.H.; Spatafora, J.W.; Kauserud, H. Fungi ahoy! Diversity on marine wooden substrata in the high North. *Fungal Ecol.* **2014**, *8*, 46–58. [CrossRef]

47. Bills, G.F.; Christensen, M.; Powell, M.; Thorn, G. Saprobic soil fungi. In *Biodiversity of Fungi, Inventory and Monitoring Methods*; Mueller, G., Bills, G.F., Foster, M., Eds.; Elsevier Academic Press: San Diego, CA, USA, 2004; pp. 271–302.

48. Collado, J.; Platas, G.; Paulus, B.; Bills, G.F. High-throughput culturing of fungi from plant litter by a dilution-to-extinction technique. *FEMS Microbiol. Ecol.* **2007**, *60*, 521–533. [CrossRef]

49. Unterseher, M.; Schnittler, M. Dilution-to-extinction cultivation of leaf-inhabiting endophytic fungi in beech (*Fagus sylvatica* L.)—Different cultivation techniques influence fungal biodiversity assessment. *Mycol. Res.* **2009**, *113*, 645–654. [CrossRef]

50. Shrestha, P.; Szaro, T.M.; Bruns, T.D.; Taylor, J.W. Systematic search for cultivatable fungi that best deconstruct cell walls of *Miscanthus* and sugarcane in the field. *Appl. Environ. Microbiol.* **2011**, *77*, 5490–5504. [CrossRef] [PubMed]

51. Jones, E.B.G. Marine fungi: Some factors influencing biodiversity. *Fungal Divers.* **2000**, *4*, 53–73.

52. Pang, K.-L.; Chow, R.; Chan, C.; Vrijmoed, L. Diversity and physiology of marine lignicolous fungi in Arctic waters: A preliminary account. *Polar Res.* **2011**, *30*, 5859–5863. [CrossRef]

53. Hyde, K.D.; Jones, E.B.G. Introduction to fungal succession. *Fungal Divers.* **2002**, *10*, 1–4.

54. Shearer, C.A. Fungi of the Chesapeake bay and its tributaries. III. The distribution of wood-inhabiting ascomycetes and fungi imperfecti of the Patuxent river. *Am. J. Bot.* **1972**, *59*, 961–969. [CrossRef]

55. Lamore, B.J.; Goos, R.D. Wood-inhabiting fungi of a freshwater stream in Rhode Island. *Mycologia* **1978**, *70*, 1025–1034. [CrossRef]

56. Rappé, M.S.; Giovannoni, S.J. The uncultured microbial majority. *Annu. Rev. Microbiol.* **2003**, *57*, 369–394. [CrossRef] [PubMed]

57. Nichols, D.; Cahoon, N.; Trakhtenberg, E.M.; Pham, L.; Mehta, A.; Belanger, A.; Kanigan, T.; Lewis, K.; Epstein, S.S. Use of Ichip for high-throughput in situ cultivation of "uncultivable" microbial species. *Appl. Environ. Microbiol.* **2010**, *76*, 2445–2450. [CrossRef] [PubMed]

58. Ling, L.L.; Schneider, T.; Peoples, A.J.; Spoering, A.L.; Engels, I.; Conlon, B.P.; Mueller, A.; Schäberle, T.F.; Hughes, D.E.; Epstein, S.; et al. A new antibiotic kills pathogens without detectable resistance. *Nature* **2015**, *517*, 455–459. [CrossRef] [PubMed]

59. Gavrish, E.; Bollmann, A.; Epstein, S.; Lewis, K. A trap for in situ cultivation of filamentous actinobacteria. *J. Microbiol. Methods* **2008**, *72*, 257–262. [CrossRef] [PubMed]

60. Epstein, S.S.; Lewis, K.; Nichols, D.; Gavrish, E. New approaches to microbial isolation. In *Manual of Industrial Microbiology and Biotechnology*; Baltz, R.H., Demain, A.L., Davies, J.E., Bull, A.T., Junker, B., Katz, L., Lynd, L.R., Masurekar, P., Reeves, C.D., Zhao, H., Eds.; ASM Press: Washington, DC, USA, 2010; pp. 3–12.
61. Mueller, G.M.; Bills, G.F.; Foster, M.S. *Biodiversity of Fungi: Inventory and Monitoring Methods*; Elsevier Academic Press: San Diego, CA, USA, 2004.
62. Kirk, P.W. Isolation and culture of lignicolous marine fungi. *Mycologia* **1969**, *61*, 174–177. [CrossRef]
63. Nakagiri, A.; Jones, E.G. Long-term maintenance of cultures. In *Marine Mycology—A Practical Approach*; Hyde, K.D., Pointing, S.B., Eds.; Fungal Diversity Press: Hong Kong, China, 2000; pp. 62–68.
64. Hyde, K.D.; Sarma, V.V. Pictorial key to higher marine fungi. In *Marine Mycology—A Practical Approach*; Hyde, K.D., Pointing, S.B., Eds.; Fungal Diversity Press: Hong Kong, China, 2000; pp. 205–270.
65. Jones, E.B.G.; Sakayaroj, J.; Suetrong, S.; Somrithipol, S.; Pang, K.-L. Classification of marine Ascomycota, anamorphic taxa and Basidiomycota. *Fungal Divers.* **2009**, *35*, 1–187.
66. Kohlmeyer, J.; Volkmann-Kohlmeyer, B. Illustrated key to the filamentous higher marine fungi. *Bot. Mar.* **1991**, *34*, 1–61. [CrossRef]
67. Volkmann-Kohlmeyer, B.; Kohlmeyer, J. How to prepare truly permanent microscope slides. *Mycologist* **1996**, *10*, 107–108. [CrossRef]
68. Schoch, C.L.; Seifert, K.A.; Huhndorf, S.; Robert, V.; Spouge, J.L.; Levesque, C.A.; Chen, W.; Consortium, F.B. Nuclear ribosomal internal transcribed spacer (ITS) region as a universal DNA barcode marker for Fungi. *Proc. Natl. Acad. Sci. USA* **2012**, *109*, 6241–6246. [CrossRef] [PubMed]
69. Fell, J.W. Yeasts in marine environments. In *Marine Fungi and Fungal-like Organisms*; Jones, E.B.G., Pang, K.-L., Eds.; Walter de Gruyter: Berlin, Germany, 2012; pp. 91–102.
70. Overy, D.P.; Correa, H.; Roullier, C.; Chi, W.-C.; Pang, K.-L.; Rateb, M.; Ebel, R.; Shang, Z.; Capon, R.; Bills, G.F.; et al. Does osmotic stress affect natural product expression in fungi? *Mar. Drugs* **2017**, *15*. [CrossRef] [PubMed]
71. Campbell, J.; Shearer, C.A.; Mitchell, J.I.; Eaton, R.A. Corollospora revisited: A molecular approach. In *Marine Mycology—A Practical Approach*; Hyde, K.D., Pointing, S.B., Eds.; Fungal Diversity Press: Hong Kong, China, 2000; pp. 15–33.
72. Schoch, C.L.; Sung, G.-H.; Volkmann-Kohlmeyer, B.; Kohlmeyer, J.; Spatafora, J.W. Marine fungal lineages in the Hypocreomycetidae. *Mycol. Res.* **2007**, *111*, 154–162. [CrossRef] [PubMed]
73. Schoch, C.L.; Sung, G.-H.; López-Giráldez, F.; Townsend, J.P.; Miadlikowska, J.; Hofstetter, V.; Robbertse, B.; Matheny, P.B.; Kauff, F.; Wang, Z.; et al. The Ascomycota tree of life: A phylum-wide phylogeny clarifies the origin and evolution of fundamental reproductive and ecological traits. *Syst. Biol.* **2009**, *58*, 224–239. [CrossRef] [PubMed]
74. Zuccaro, A.; Schoch, C.L.; Spatafora, J.W.; Kohlmeyer, J.; Draeger, S.; Mitchell, J.I. Detection and identification of fungi intimately associated with the brown seaweed *Fucus serratus*. *Appl. Environ. Microbiol.* **2008**, *74*, 931–941. [CrossRef]
75. Suetrong, S.; Schoch, C.L.; Spatafora, J.W.; Kohlmeyer, J.; Volkmann-Kohlmeyer, B.; Sakayaroj, J.; Phongpaichit, S.; Tanaka, K.; Hirayama, K.; Jones, E.B.G. Molecular systematics of the marine *Dothideomycetes*. *Stud. Mycol.* **2009**, *64*, 155–173. [CrossRef]
76. Abraham, S.; Hoang, T.; Alam, M.; Jones, E.G. Chemistry of the cytotoxic principles of the marine fungus *Lignincola laevis*. *Pure Appl. Chem.* **1994**, *66*, 2391–2394. [CrossRef]
77. Lin, Y.; Wu, X.; Deng, Z.; Wang, J.; Zhou, S.; Vrijmoed, L.; Jones, E.G. The metabolites of the mangrove fungus *Verruculina enalia* No. 2606 from a salt lake in the Bahamas. *Phytochemistry* **2002**, *59*, 469–471. [CrossRef]
78. Poch, G.K.; Gloer, J.B. Obionin A: A new polyketide metabolite form the marine fungus *Leptosphaeria obiones*. *Tetrahedron Lett.* **1989**, *30*, 3483–3486. [CrossRef]
79. Poch, G.K.; Gloer, J.B. Helicascolides A and B: New lactones from the marine fungus *Helicascus kanaloanus*. *J. Nat. Prod.* **1989**, *52*, 257–260. [CrossRef] [PubMed]

Article

Diversity and Ecology of Marine Algicolous *Arthrinium* Species as a Source of Bioactive Natural Products

Young Mok Heo [1], Kyeongwon Kim [1], Seung Mok Ryu [2], Sun Lul Kwon [1], Min Young Park [2], Ji Eun Kang [2], Joo-Hyun Hong [1], Young Woon Lim [3], Changmu Kim [4], Beom Seok Kim [2], Dongho Lee [2,*] and Jae-Jin Kim [1,*]

[1] Division of Environmental Science & Ecological Engineering, College of Life Sciences & Biotechnology, Korea University, Seoul 02841, Korea; hym011@korea.ac.kr (Y.M.H.); rudndjs@korea.ac.kr (K.K.); sun-lul@korea.ac.kr (S.L.K.); dress8@korea.ac.kr (J.-H.H.)
[2] Division of Biotechnology, College of Life Sciences & Biotechnology, Korea University, Seoul 02841, Korea; mogijjang@korea.ac.kr (S.M.R.); min0@korea.ac.kr (M.Y.P.); heyyo9725@korea.ac.kr (J.E.K.); biskim@korea.ac.kr (B.S.K.)
[3] School of Biological Sciences and Institute of Microbiology, Seoul National University, Seoul 08826, Korea; ywlim@snu.ac.kr
[4] Microorganism Resources Division, National Institute of Biological Resources, Incheon 22689, Korea; snubull@korea.kr
* Correspondence: dongholee@korea.ac.kr (D.L.); jae-jinkim@korea.ac.kr (J.-J.K.); Tel.: +82-2-3290-3017 (D.L.); +82-2-3290-3049 (J.-J.K.); Fax: +82-2-953-0737 (D.L.); +82-2-3290-9753 (J.-J.K.)

Received: 8 November 2018; Accepted: 10 December 2018; Published: 14 December 2018

Abstract: In our previous study, all *Arthrinium* isolates from *Sargassum* sp. showed high bioactivities, but studies on marine *Arthrinium* spp. are insufficient. In this study, a phylogenetic analysis of 28 *Arthrinium* isolates from seaweeds and egg masses of *Arctoscopus japonicus* was conducted using internal transcribed spacers, nuclear large subunit rDNA, β-tubulin, and translation elongation factor region sequences, and their bioactivities were investigated. They were analyzed as 15 species, and 11 of them were found to be new species. Most of the extracts exhibited radical-scavenging activity, and some showed antifungal activities, tyrosinase inhibition, and quorum sensing inhibition. It was implied that marine algicolous *Arthrinium* spp. support the regulation of reactive oxygen species in symbiotic algae and protect against pathogens and bacterial biofilm formation. The antioxidant from *Arthrinium* sp. 10 KUC21332 was separated by bioassay-guided isolation and identified to be gentisyl alcohol, and the antioxidant of *Arthrinium saccharicola* KUC21221 was identical. These results demonstrate that many unexploited *Arthrinium* species still exist in marine environments and that they are a great source of bioactive compounds.

Keywords: antioxidant; biological control; ecological role; gentisyl alcohol; multi-gene phylogeny; tyrosinase inhibition

1. Introduction

Among a number of marine fungi isolated from a brown alga *Sargassum* sp., *Arthrinium* spp. showed high levels of cellulolytic enzyme productivity, radical-scavenging, or antifungal activities [1]. To date, however, studies on marine *Arthrinium* spp. are insufficient in terms of reliable phylogenetic analysis, physiological activity, bioactive secondary metabolites, and ecology.

The genus *Arthrinium* Kunze (sexual morph *Apiospora*; Ellis 1971 [2], Apiosporaceae) was reported as an endophyte in plant and ecologically-diverse species occurring in various habitats in both terrestrial and marine environments [3,4]. The genus *Arthrinium* has numerous broad synonyms [5].

Cordella is a potential synonym for *Arthrinium* and distinguished primarily by the existence of setae. *Pteroconium*, with the uncertain status of its generic name, has been regarded as an individual genus separate from *Arthrinium*, although *Apiospora* is the sexual morph of both *Pteroconium* and *Arthrinium* [2,5,6]. Currently, 80 *Arthrinium* species are reported in Index Fungorum (2018), but most *Arthrinium* taxa do not have enough sequence data and lack a detailed description of their morphological characteristics. Recently, the genus *Arthrinium* was re-examined by phylogenetic analysis of 16 species, which suggested that it is necessary to use β-tubulin (TUB) and translation elongation factor (TEF) regions to resolve species complexes [4]. Recently, Wang et al. (2018) published eight new *Arthrinium* spp. using three combined loci and multi-gene phylogenetic analysis with high bootstrap supports [7].

Marine-derived fungi have been suggested as a novel source of bioactive secondary metabolites, which can be applied in the pharmaceutical and cosmetic industry [8]. This suggestion was due to cephalosporin C production by *Acremonium chrysogenum* being reported in 1961 as the first bioactive metabolite from marine fungi, but novel natural compounds from marine fungi were less explored in comparison with compounds from terrestrial fungi [9]. Marine natural compounds have provided sufficient interest to the pharmaceutical industry due to their unique chemical properties [10]. In particular, marine fungal endophytes have been recognized as potential producers for bio-based commercial products [11]. Marine *Arthrinium* spp. have been reported as an endo-symbiont of marine algae, especially brown algae. They have been discovered to produce natural compounds such as arthrinone, arthrichitin, terpestacin, (1R,2S,3aS,8aR)-3a,6-dimethyl-1-(propan-2-yl)-1,2,3,3a,4,7,8,8a-octahydroazulene-1, 2-diol (CAF-603), norlichexanthone, myrocins, libertellenones, spiroarthrinols, and griseofulvin derivatives [12–15]. However, a limited number of bioactive compounds were reported: arthpyrones F–I and apiosporamide (antibacterial) [16]; arthrinins A–D (antitumoral and antiproliferative) [13]; decarboxyhydroxycitrinone, myrocin A, libertellenone C, and cytochalasin E (antiangiogenic); arthone C and 2,3,4,6,8-pentahydroxy-1-methylxanthone (antioxidant) [17]; and cytochalasin K and 10-phenyl-[12]-cytochalasin Z_{16} (cytotoxic) [18]. In this regard, it was suggested that marine *Arthrinium* spp. have great potential to produce various bioactive compounds that can be the ingredients of a wide range of bioagents, many of which have not been discovered yet. In this study, the biological activities of marine *Arthrinium* spp. extracts were investigated to evaluate the value of marine *Arthrinium* spp. as a source of bioactive compounds and to infer their ecological role in symbiotic relationships with marine algae.

The aims of this study were (1) to identify and phylogenetically analyze marine-derived *Arthrinium* spp., (2) to evaluate the pharmaceutical value of marine *Arthrinium* spp. as a source of bioactive compounds, (3) to evaluate the ecological role in symbiosis with brown algae, and (4) to isolate and structurally identify a bioactive compound from marine *Arthrinium* species.

2. Results and Discussion

2.1. Identification and Multi-Gene Phylogeny

A total of 28 marine-derived *Arthrinium* strains were isolated from an unknown seaweed, *Sargassum* sp., *Agarum cribrosum*, and egg masses of *Arctoscopus japonicus* (Table 1). For the phylogenetic analysis, we constructed a phylogenetic tree using combined datasets from internal transcribed spacers (ITS), partial nuclear large subunit rDNA (LSU), TUB, and TEF 1-alpha (EF-1α) region sequences containing 45 taxa and 680, 559, 590, and 534 nucleotide characters, respectively (Figure 1). As the results of the MrModeltest with the Akaike information criterion (AIC), a general time reversible (GTR) + proportion of invariable sites (I) + gamma distribution (G) model was chosen for LSU and TUB, and the symmetrical model (SYM) + I + G model and Hasegawa-Kishino-Yano (HKY) + I + G model were chosen for ITS and EF-1α, respectively. The form of the tree demonstrated that all the species were clustered with high posterior probabilities (95–100%). As a result, the 28 *Arthrinium* strains were divided into 15 species, including 11 novel species candidates. All new species will be reported in the near future.

Table 1. General information for the 28 marine–derived *Arthrinium* spp. and antioxidant activity of the fungal extracts.

Fungal Identity	Strain ID	Isolation Source	GenBank Accession Number				Antioxidant Activity (IC$_{50}$, μg/mL)	
			ITS	LSU	TUB	EF-1α	ABTS [a]	DPPH [b]
A. arundinis	KUC21229	*Sargassum fulvellum*	KT207747	MH498470	MH498512	MH544684	40.55	>1000
	KUC21261	*Sargassum fulvellum*	KT207779	MH498469	MH498511	MH544683	52.33	>1000
	KUC21335	Unknown seaweed	MH498552	N.D.	MH498510	N.D.	64.89	N.D.
	KUC21337	Beach sand	MH498551	MH544659	MH498509	MH544682	>100	>1000
A. marii	KUC21338	Unknown seaweed	MH498549	MH498467	MH498507	MH544681	74.77	N.D.
A. sacchari	KUC21340	Egg masses of *Arctoscopus japonicus*	MH498548	MH498466	MH498506	MH544680	17.45	601.34
A. saccharicola	KUC21221	*Sargassum fulvellum*	KT207737	KT207687	KT207637	MH544679	14.09	59.19
	KUC21341	Egg masses of *Arctoscopus japonicus*	MH498547	MH498465	MH498505	N.D.	>100	N.D.
	KUC21342	Egg masses of *Arctoscopus japonicus*	MH498546	MH498464	MH498504	N.D.	58.27	N.D.
	KUC21343	Egg masses of *Arctoscopus japonicus*	MH498545	MH498463	MH498503	MH544678	38.81	744.32
Arthrinium sp. 1	KUC21228	*Sargassum fulvellum*	KT207746	KT207696	KT207644	MH544677	55.29	N.D.
	KUC21232	*Sargassum fulvellum*	KT207750	KT207700	KT207648	MH544676	>100	>1000
	KUC21282	*Sargassum fulvellum*	MH498544	MH498462	MH498502	MH544675	45.37	>1000
	KUC21284	*Sargassum fulvellum*	MF615228	MF615215	MF615233	MH544674	100.04	>1000
	KUC21334	Egg masses of *Arctoscopus japonicus*	MH498543	MH544661	MH498501	MH544673	50.89	>1000
Arthrinium sp. 2	KUC21220	*Sargassum fulvellum*	KT207736	KT207786	KT207636	MH544672	23.23	>1000
	KUC21279	*Sargassum fulvellum*	MF615229	MF615216	MF615234	MH544671	41.71	>1000
	KUC21280	*Sargassum fulvellum*	MH498542	MH544660	MH498500	N.D.	61.43	>1000
Arthrinium sp. 3	KUC21327	Egg masses of *Arctoscopus japonicus*	MH498541	MH498461	MH498499	MH544670	10.32	>1000
Arthrinium sp. 4	KUC21328	Unknown seaweed	MH498538	MH498458	MH498496	MH544669	19.22	>1000
Arthrinium sp. 5	KUC21288	Unknown seaweed	MF615230	MF615217	MF615235	MH544668	50.32	>1000
	KUC21289	Unknown seaweed	MF615226	MF615213	MF615231	MH544668	61.63	>1000
Arthrinium sp. 6	KUC21321	Unknown seaweed	MH498533	MH498453	MH498491	N.D.	46.98	>1000
Arthrinium sp. 7	KUC21329	Egg masses of *Arctoscopus japonicus*	MH498531	MH498451	MH498489	MH544666	39.05	>1000
Arthrinium sp. 8	KUC21330	Egg masses of *Arctoscopus japonicus*	MH498530	MH498450	MH498488	MH544665	72.56	>1000
Arthrinium sp. 10	KUC21332	Egg masses of *Arctoscopus japonicus*	MH498524	MH498444	MH498482	MH544664	14.87	73.22
Arthrinium sp. 11	KUC21333	*Agarum cribrosum*	MH498520	MH498440	MH498478	MH544663	32.62	>1000
Arthrinium sp. 12	KUC21322	Unknown seaweed	MH498515	MH498435	MH498473	MH544662	>100	>1000
Ascorbic acid *							13.70	6.80

N.D. means no data or not detected. [a] ABTS, 2,2'-azino-bis-3-ethylbenzothiazoline-6-sulfonic acid; [b] DPPH, 2,2-diphenyl-1-picrylhydrazyl; * positive control for antioxidant activity.

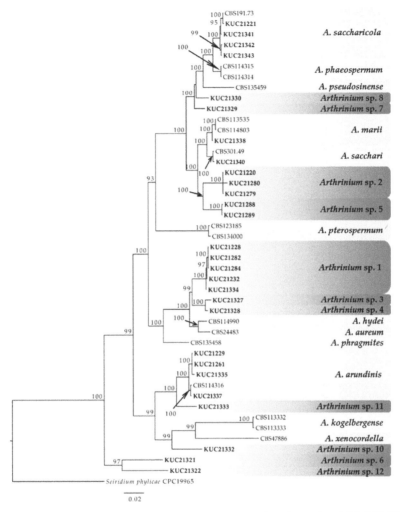

Figure 1. The phylogenetic tree of the *Arthrinium* species based on the combined ITS, LSU, TUB, and EF-1α sequence alignment from Bayesian analysis using MrBayes v. 3.2.1. The dataset was created from 45 taxa and 2363 characters. The fungal cultures examined in this study are boldfaced. The number above the branch indicates the posterior probability values ≥70. The scale bar indicates nucleotide substitutions per position.

2.2. Marine Habitat of Arthrinium

Most of the *Arthrinium* species isolated from each substrate were identified to be different (Table 1). Seven different *Arthrinium* species were isolated from the egg masses of *A. japonicus*. The two known species were *Arthrinium saccharicola* and *A. sacchari*, and the other five strains were considered novel *Arthrinium* candidates. Interestingly, *A. saccharicola*, *A. arundinis*, and *Arthrinium* sp. 1 were found in both *Sargassum fulvellum* and the egg masses of *A. japonicus* [19]. *A. japonicus* spawns the largest number of eggs on *Sargassum fulvellum*, one of the major endophytic hosts of *Arthrinium* spp. [20]. It was strongly speculated that the separation of the same *Arthrinium* species from both substrates was based on this fact. The conidia of the algicolous *Arthrinium* spp. can easily be transferred to the egg masses of *A. japonicus*. In another study, the fungal diversity of the *A. japonicus* egg masses were investigated, and eight *Arthrinium* species were founded: *A. saccharicola*, *A. sacchari*, *A. phaeospermum*, *A. arundinis*, and *A. rasikravindri*, as well as three unknown *Arthrinium* species [19]. According to Park et al. (2018), *Arthrinium* accounts for a small portion of the strains isolated from egg masses of *A. japonicus*. Nonetheless, the fact that there were many new species candidates suggests that various unexplored *Arthrinium* species are present in the marine environment.

A. saccharicola, known as a plant pathogenic fungus, has been isolated from living and dead culms of *Phragmites australis* and even from the air in the Netherlands and France and from seawater in mangrove habitats in Hong Kong [4,21]. This suggested that *A. saccharicola* has low host specificity. Additionally, *A. arundinis*, *A. marii*, and *A. sacchari* have been isolated from more than one substrate in another study [7]. *A. arundinis* has been isolated from the leaf of *Hordeum vulgare* and living leaves of *Fagus sylvatica* in Iran and Switzerland [4]. Additionally, in a recent study, 19 strains of *A. arundinis* were isolated from nine different hosts in China [7]. Taxonomical work on *Arthrinium* species isolated from marine ecosystems has been relatively lacking compared to work on those isolated from terrestrial ecosystems. Thus, more taxonomic and systematic analyses of marine-derived *Arthrinium* spp. are needed.

2.3. Biological Activities of Marine Arthrinium spp.

2.3.1. Antioxidant Activity

Since *Arthrinium* spp. showed high antioxidant activity, other marine-derived *Arthrinium* species were also expected to do so [1]. The results showed that almost every *Arthrinium* strain showed high radical-scavenging activity on the assay using 2.2′-azino-bis-3-ethylbenzothiazoline-6-sulfonic acid (ABTS) radicals as a substrate (Table 1). In particular, the crude extract of *Arthrinium* sp. 3 KUC21327 exhibited even higher ABTS radical-scavenging activity than ascorbic acid. Reactive oxygen species (ROS) are responsible for many kinds of diseases and aging because they damage biomolecules, such as DNA and protein, through oxidative chain reactions [22]. This suggests that the symbiotic relationship between *Arthrinium* spp. and their major host, seaweeds, may be based on the regulation of the ROS-defense system. Many seaweeds have been used as functional food and cosmetic ingredients because of their great antioxidant ability [23,24]. It was speculated that these algicolous fungi might contribute to the antioxidant ability of seaweeds.

Researchers have discovered many kinds of antioxidants, which can scavenge free-radicals generated from ROS, and filamentous fungi were one of the main natural sources of antioxidants. Filamentous fungi produce a wide range of secondary metabolites, and many polyphenolic compounds have been reported to exhibit high radical-scavenging abilities against ROS and free radicals [25]. In the case of marine *Arthrinium* spp., arthone C and 2,3,4,6,8-pentahydroxy-1-methylxanthone were reported as antioxidants [17]. Our ABTS assay results imply that marine-derived *Arthrinium* spp. commonly produce certain kinds of antioxidant compounds. Among them, *A. sacchari* KUC21340, *A. saccharicola* KUC21221, *A. saccharicola* KUC21343, and *Arthrinium* sp. 10 KUC21332 exhibited high radical-scavenging activity against the 2,2-diphenyl-1-picrylhydrazyl (DPPH) radical. Because the reaction media of ABTS and DPPH assays differ (PBS buffer and 80% MeOH, respectively), the measured activity is affected by the solubility of potential antioxidant compounds for each solvent [26]. These four strains produce antioxidant compounds that are less hydrophilic than those of the other strains and remain active in MeOH. In particular, the IC_{50} values of *A. saccharicola* KUC21221 and *Arthrinium* sp. 10 KUC21332 were significantly lower than those of the other species, implying that these species have high antioxidative capacities. The high radical-scavenging activity of *A. saccharicola* KUC21221 extract was already reported [1]. Among the two candidates, *Arthrinium* sp. 10 was selected in consideration of the novelty of this fungal species, its antioxidant compound was separated, and the chemical structure was identified.

2.3.2. Antifungal Activity

In our previous study, all six *Arthrinium* sp. isolated from *Sargassum* sp., a brown alga, exhibited high antifungal activities against terrestrial plant pathogens [1]. As *Arthrinium* spp. are one of the key endophytes of seaweed, some *Arthrinium* species were expected to have symbiotic relationships in immune mechanisms by producing antibiotics against other marine microorganisms harmful to the algae. As a result, the extracts of three *A. saccharicola* (KUC21221, KUC21341, and KUC21343) and *Arthrinium* sp. 2 KUC21220 showed inhibitory effects against mycelial growth of the test fungus *Asteromyces cruciatus*, an alginate-degrading fungus (Table S2). Alginate is a major constituent of brown algae and plays an important role in brown algae by forming the structure of the algal biomass and physically protecting algae from pathogens [27]. Some marine fungi have been reported to be able to degrade alginate: *Asteromyces cruciatus*, *Corollospora intermedia*, *Dendryphiella arenaria*, and *D. salina* [28]. In particular, *A. cruciatus* is one of the major marine colonizers ubiquitous in the marine environment [29]. This fungus can be regarded as a potentially harmful microorganism to brown algae, as it can colonize the entire algal biomass and breakdown alginate enzymatically when the defense system of algae is weakened. Decomposition of the biomaterial leads to the collapse of the defense system.

In particular, the two *A. saccharicola* (KUC21341 and KUC21342) extracts had the highest antifungal activities as they could inhibit the growth of *A. cruciatus* at a concentration of 100 µg/mL. Since they were identified as the same species, the antifungal secondary metabolites are highly likely to be identical. This result implied that the two *Arthrinium* species, especially *A. saccharicola*, may have a symbiotic relationship with brown algae that assists its defense system against other marine fungi. Furthermore, it was strongly speculated that their antifungal metabolites can affect other pathogenic fungi, considering that six marine-derived *Arthrinium* strains (*A. arundinis* KUC21229 and KUC21261, *A. saccharicola* KUC21221, *Arthrinium* sp. 1 KUC21228 and KUC21232, and *Arthrinium* sp. 2 KUC21220) showed antifungal activity against several terrestrial plant pathogenic fungi [1]. Considering that most antimicrobials have low specificity, it was thought that they might have evolved to synthesize antifungal compounds not only for themselves, but also for their symbiotic partner, marine algae. Plants are protected by cocktails of moderately-active compounds, such as essential oils, and synergy with antimicrobials lowers the minimum effective concentration of individual antimicrobial compound to prevent the emergence of resistant pathogens [30]. Similarly, it was suggested that the broad spectrum antifungal potential of these marine algicolous *Arthrinium* spp. may contribute to improving the fitness of the algal hosts by providing a broad spectrum of defense that does not trigger the emergence of resistant pathogens. It is necessary to study the interaction between antimicrobials of *Arthrinium* spp. and secondary metabolites of the algal hosts.

2.3.3. Tyrosinase Inhibition Activity

A total of eight *Arthrinium* extracts exhibited inhibitory activity against tyrosinase, and *A. sacchari* KUC21340, *Arthrinium* sp. 2 KUC21279, and *Arthrinium* sp. 7 KUC21329 were the best producers of tyrosinase inhibitors (Table S2). Tyrosinase catalyzes the oxidation of not only tyrosine, but also L-3,4-dihydroxyphenylalanine (L-DOPA), a reactive intermediate compound of melanin, to dopaquinone, which eventually converts it to pheomelanin [31]. Tyrosinase inhibitors suppress the formation of melanin in human skin and prevent the darkening of the skin tone, and researchers have discovered several inhibitors of fungal origin, such as protocatechualdehyde from *Phellinus linteus*, 6-n-pentyl-α-pyrone from *Myrothecium* sp., and kojic acid, the most widely-studied tyrosinase inhibitor, from various fungi, especially *Aspergillus oryzae* [32–34]. These tyrosinase inhibitors have various mechanisms, and many antioxidant compounds have been reported to exhibit tyrosinase inhibition activity by suppressing melanin biosynthesis by scavenging reactive quinone products [31]. The extracts of *A. sacchari* KUC21340, *Arthrinium* sp. 2 KUC21279, and *Arthrinium* sp. 7 KUC21329 exhibited both tyrosinase inhibitory and radical-scavenging activity, and it can be interpreted that the tyrosinase inhibitors can be identical to their antioxidant compounds. This result indicates that the three *Arthrinium* species can produce tyrosinase inhibitors with antioxidant activity that regulates melanin biosynthesis by scavenging reactive quinone products.

2.3.4. Quorum Sensing Inhibition Activity

The extracts of *Arthrinium* sp. 1 KUC21228 and KUC21232 and *Arthrinium* sp. 6 KUC21321 inhibited the production of violacein by *C. violaceum* CV026 in the presence of *N*-(3-oxo-hexanoyl)-L-homoserine lactone (3-oxo-C6-HSL), an *N*-acyl homoserine lactone (AHL), which is one of the most widely-studied quorum sensing (QS) molecules (Figure 2). In particular, *Arthrinium* sp. 1 KUC21228 extract showed a high inhibitory activity even comparable to that of the positive control, which is a pure compound (Figure 2A). QS is responsible for the formation of bacterial biofilms and the expression of virulence genes, and bacterial biofilms have been reported to be approximately 1000-times more resistant to antibiotics than their planktonic counterparts [35]. Some fungi produce quorum-quenching compounds such as patulin and penicillic acid from *Penicillium* spp., which can suppress bacterial biofilm formation [36]. All four *Arthrinium* strains that showed QS inhibition activity were isolated from seaweeds that are well known to produce various quorum sensing inhibitors (QSIs), namely bromoperoxidase from a brown alga *Laminaria digitate* and floridoside, bentonicine, and isoethionic acid from a Korean red alga *Ahnfeltiopsis flabelliformis* [37,38].

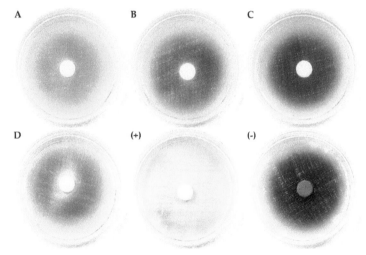

Figure 2. Quorum sensing inhibitory activity of extracts from four marine algicolous *Arthrinium* strains. (**A**) *Arthrinium* sp. 1 KUC21228; (**B**) *Arthrinium* sp. 1 KUC21232; (**C**) *Arthrinium* sp. 5 KUC21289; (**D**) *Arthrinium* sp. 6 KUC21321; (+) positive control (piericidin A); (−) negative control (DMSO).

In fact, many filamentous fungi associated with the plant rhizosphere have been reported to degrade 3-oxo-C6-HSL enzymatically [39]. Therefore, it was suggested that these seaweed-endophytic *Arthrinium* species can produce QSIs that are expected to be lactonase or other kinds of AHL-degrading enzymes, which degrade 3-oxo-C6-HSL and assist in resistance to bacterial biofilms in seaweeds. Otherwise, it could be norlichexanthone, a known metabolite of *Arthrinium* sp., reported to inhibit the QS mechanism in *Staphylococcus aureus*, a human pathogen [40]. Considering that arthpyrones F-I and apiosporamide were reported as antibacterial compounds from marine *Arthrinium* sp., some marine *Arthrinium* species may directly inhibit seaweed-pathogenic bacteria [16]. The three new seaweed-endophytic *Arthrinium* strains were revealed to be potential QSI producers, which makes them valuable as a source of biological compounds that inhibit the bacterial biofilm, which causes huge economic losses in many areas, such as food, aquaculture, wastewater treatment, and the shipping industry [41,42]. The selected fungi will be subjected to an in vitro experiment as a subsequent screening by quantifying their activity.

2.4. Antioxidant Compound from Arthrinium sp. 10

The extract of *Arthrinium* sp. 10 KUC21332, a new species candidate with the highest radical-scavenging activity, was selected to identify the active compound. A bioassay-guided isolation of this extract afforded gentisyl alcohol (**1**) (Figure 3). The chemical structure was identified using spectroscopic data and a comparison with literature data [43]. The ^{1}H and ^{13}C NMR spectra, LC chromatogram, UV spectrum, and MS spectra of Compound **1** are shown in the Supplementary Materials (Figures S1–S4). In addition, the results of comparative analysis of *A. saccharicola* KUC21221 extract, *Arthrinium* sp. 10 KUC21332 extract, and gentisyl alcohol using ultra-performance liquid chromatography (UPLC) with a photodiode array (PDA) detector confirmed that the antioxidant compound of *A. saccharicola* KUC21221 is the same as that of *Arthrinium* sp. 10 KUC21332, that is gentisyl alcohol (Figure S5).

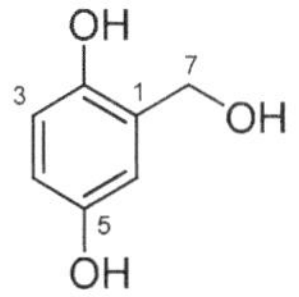

Figure 3. Structure of Compound **1**.

The radical-scavenging activity of Compound **1** was evaluated using the ABTS and DPPH assays. As described in Table 2, it exhibited higher activities than the positive control (ascorbic acid), which corresponds to the result of a previous report [44]. Gentisyl alcohol has also been reported to have proangiogenic, caspase inhibition, histone deacetylases inhibition, antibacterial (against methicillin-resistant *Staphylococcus aureus*), antifungal (against *Colletotrichum gloeosporioides*), antileishmanial (against *Leishmania donovani*), and cytotoxic (against human breast and colon cancer cell lines, MCF-7 and HT-29, respectively, and against sea urchin *Strongylocentrotus intermedius*) activity [45–53]. This is the first report of the production of gentisyl alcohol from the genus *Arthrinium*.

Table 2. ABTS and DPPH radical-scavenging activity of Compound **1** isolated from *Arthrinium* sp. 10 KUC21332.

Compound	Antioxidant Activity (IC$_{50}$, μM)	
	ABTS [a]	DPPH [b]
Gentisyl alcohol (**1**)	26.43	28.74
Ascorbic acid *	77.79	38.61

[a] ABTS, 2.2′-azino-bis-3-ethylbenzothiazoline-6-sulfonic acid; [b] DPPH, 2,2-diphenyl-1-picrylhydrazyl; * positive control.

3. Materials and Methods

3.1. Microorganisms

3.1.1. Fungal Resources

Marine-derived *Arthrinium* spp. used in this study were obtained from the Korea University Culture (KUC) collection and Marine Fungal Resource Bank (MFRB) at Seoul National University as a marine bioresource bank of Korea by the Ministry of Oceans and Fisheries. *Asteromyces cruciatus* SFC20161110-M19 was obtained from MFRB to use as a target fungus of the antifungal assay.

The algicolous fungi were isolated according to the following procedure [1,19]. Marine algae and the egg masses of *A. japonicus* were washed with sterile 3.44% artificial sea water (ASW) and cut into 5-mm^2 squares. The pieces were placed on each of three culture plates: dichloran rose bengal chloramphenicol agar (Difco, Sparks, MD, USA), glucose yeast extract agar (1 g/L glucose, 0.1 g/L yeast extract, 0.5 g/L peptone, and 15 g/L agar), and Sabouraud dextrose agar (Difco, Sparks, MD, USA) supplemented with 3.44% ASW, 0.01% streptomycin, and 0.01% ampicillin to prevent bacterial growth. The plates were incubated at 25 °C for 7–15 days, and the grown fungi were transferred to a potato dextrose agar (PDA, Difco, Sparks, MD, USA) plate with 3.44% ASW periodically.

3.1.2. DNA Extraction, PCR, and Identification

Genomic DNA extraction was performed using fungal culture grown on malt extract agar by the AccuPrep Genomic DNA Extraction kit (Bioneer, Seoul, Korea). The DNA sequences of KUC21221, KUC21229, KUC21261, KUC21228, KUC21232, and KUC21220 were obtained from our previous research [1]. PCR reactions were performed with three different primer regions using the AccuPower PCR Premix Kit (Bioneer, Seoul, Korea). ITS (forward: ITS1F (5′-CTTGGTCATTTAGAGGAAGTAA-3′)) and reverse: LR3 (5′-CCGTGTTTCAAGACGGG-3′)) [54,55], LSU (forward: LR0R (5′-ACCCGCTGAACTTAAGC-3′) and reverse: LR5 (5′-TCCTGAGGGAAACTTCG-3′) or LR7 (5′-TACTACCACCAAGATCT-3′)) [56], TUB (forward: T10 (5′-CATCGAGAAGTTCGAGAAGG-3′) or Bt2a (5′-GGTAACCAAATCGGTGCTGCTTTC-3′) and reverse: T2 (5′-TAGTGACCCTTGGCCCA GTTG-3′) or Bt2b (5′-ACCCTCAGTGTAGTGACCCTTGGC-3′)) [57,58], and EF-1α (forward: EF1-728F (5′-GGA(G/A)GTACCAGT(G/C)ATCATGTT-3′) and reverse: EF2 (5′-GGA(G/A)GTACCAGT(G/C) ATCATGTT-3′)) [59,60] were amplified under the temperature cycling parameters as follows. For ITS and LSU, 95 °C for 4 min, followed by 30 cycles of 95 °C for 30 s, 55 °C for 30 s, and 72 °C for 30 s; an elongation step of 72 °C for 5 min was performed at the end. For TUB, 95 °C for 5 min, followed by 30 cycles of 95 °C for 35 s, 55 or 56 °C for 50 s, and 72 °C for 2 min; an elongation step was performed at 72 °C for 7 min. For EF-1α, 94 °C for 2 min, followed by 29 cycles of 93 °C for 30 s, 55 or 56 °C for 30 s, and 72 °C for 1 minute; an elongation step was performed at 72 °C for 10 min. DNA sequencing was executed by Macrogen (Seoul, Korea) using the Sanger method with a 3730xl DNA Analyzer (Life technology, Carlsbad, CA, USA). The obtained DNA sequences were submitted to the GenBank, and the accession numbers are presented in Table 1.

3.1.3. Phylogenetic Analysis

The obtained DNA sequences with reference sequences (Table S1) obtained from GenBank using a BLAST search were proofread, aligned using MAFFT 7.397, and modified manually using MacClade v4.08 (http://macclade.org) [61,62]. They were tested by MrModeltest v2.3 (https://www.softpedia. com/get/Science-CAD/MrModeltest.shtml) with default options using the AIC [63]. Bayesian analysis was performed by MrBayes v3.2.1 (http://nbisweden.github.io/MrBayes) [64]. Two operations were performed and contained 1,000,000 generations. For each operation, the result of every 100th generation was sampled. Among them, the first 25% of the trees was removed, and the last 75% was selected. A phylogenetic tree was constructed according to the 50% majority-rule, and tree reliability was confirmed by the posterior probability.

3.2. *Preparation of Fungal Extracts*

All of the fungal species were precultured on 20-mL Petri dishes containing PDA at 25 °C until mycelia covered the dishes. After the preculture period, three agar plugs with mycelia were transferred to Petri dishes (150 × 20 mm) containing 50 mL of PDA and incubated at 25 °C for seven days in darkness. The fungal cultures were incubated in triplicate. To obtain fungal extracts, the solid media with mycelia were extracted with 200 mL of methanol for 24 h, followed by filtration with Whatman No.1 filter paper. Filtrates were dried at 37 °C under a vacuum and cooled at 4 °C during circulation. The dried residues were redissolved in a half and half percentage of water-ethyl acetate solution. After six hours, the portion of ethyl acetate was collected and evaporated under the same temperature condition mentioned above. The dried extracts were maintained in the 20-mL scintillation vials at 4 °C until use. To obtain a large amount of crude extracts for separating the bioactive compound, a culture and extraction were carried out using the same method in 200 Petri dishes.

3.3. Biological Assays

3.3.1. Antioxidant Assay

ABTS Radical-Scavenging Assay

ABTS (Sigma-Aldrich, Inc., St. Louis, MO, USA) was dissolved in phosphate-buffered saline (PBS, pH 7.4) to 7 mM. Then, the ABTS solution was mixed with potassium persulfate solution that was dissolved in PBS to 2.45 mM. The mixture was stored in the dark at room temperature for 24 h. After the ABTS$^{\bullet+}$ radical was formed, the ABTS solution was diluted with PBS to an absorbance of 0.70 ($\pm$0.02) at a wavelength of 734 nm. Then, 990 μL of ABTS radical solution and 10 μL of fungal extract (10 mg/mL in DMSO) were reacted in a cuvette. The absorbance was measured at 734 nm after 6 min using a spectrophotometer (Sunrise^{TM}, Tecan, Männedorf, Switzerland).

DPPH Radical-Scavenging Assay

DPPH (Sigma-Aldrich Inc., St. Louis, MO, USA) was dissolved in 80% methanol at 150 μM. The 200 μL of DPPH solution and 22 μL of the fungal extracts (10 mg/mL) was mixed in each well in a 96-well plate. The plate was allowed to reach a steady state for 30 min at room temperature. The absorbance was measured at 540 nm.

3.3.2. Antifungal Assay

Antifungal activity was determined in a 96-well plate using the microtiter broth dilution method [65]. Twenty-five microliters of spore suspensions (4×10^5 conidia/mL) of *Asteromyces cruciatus* SFC20161110-M19 were added to each well containing 25 μL of 4× potato dextrose broth and 49 μL of D.W. Finally, the fungal extracts were added to a final concentration of 100 μg/mL. The inoculated 96-well plates were incubated at 25 °C, for three days. To determine the minimum inhibitory concentration (MIC), the extracts with lower concentrations (50, 25, 12.5, 6.5 μg/mL) were tested.

3.3.3. Tyrosinase Inhibition Assay

A tyrosinase inhibition assay was conducted based on the method described by Lai et al. (2009) [66]. Forty microliters of fungal extract (2.5 mg/mL in 50% DMSO), 70 μL of 0.1 M potassium phosphate buffer (pH 6.8), and 30 μL of tyrosinase from mushroom (0.02 mg/mL in the buffer; Sigma Aldrich, St. Louis, MO, USA) were mixed in a 96-well plate. The mixture was heated to 30 °C for five minute, and 100 μL of 2.5 mM L-DOPA (Sigma Aldrich, St. Louis, MO, USA) was added to the well. After 30 min, the reaction was terminated by putting the plate in ice, and the absorbance was measured at 492 nm. Kojic acid (Sigma Aldrich, St. Louis, MO, USA) was used as a positive control, and a mixture without L-DOPA was regarded as a blank.

3.3.4. Quorum Sensing Inhibition Assay

QSI screening was performed based on the inhibition of violacein production by *Chromobacterium violaceum* CV026 strains under culture conditions supplemented with an exogenous QS molecule, 3-oxo-C6-HSL [67]. The *C. violaceum* CV026 cultured overnight was diluted with LB medium (5 g of yeast extract, 10 g of tryptone, and 5 g of NaCl in a liter of D.W.) to an OD$_{600nm}$ of 0.1, and 97 μL of the diluted culture were added to each well of a 96-well plate. One microliter of 3-oxo-C6-HSL (final concentration of 10 μM; Sigma-Aldrich, St. Louis, MO, USA) and two microliters of fungal extract (diluted to 10 mg/mL in DMSO) was added to each well. As a negative control and positive control, DMSO and 100 nM (*Z*)-4-bromo-5-(bromomethylene)-2(5*H*)-furanone (Furanone C-30; Sigma-Aldrich,

St. Louis, MO, USA) were used, respectively. After 16 h at 28 °C, 100 µL of DMSO was added to each well and vigorously shaken for an hour at room temperature to determine the production of violacein.

The selected extracts were further examined using a paper disc method. Fifteen microliters of the diluted culture of *C. violaceum* CV026 was spread on an LB agar plate, and a paper disc (diameter 8 mm; Advantec, Tokyo, Japan) impregnated with 40 µL of extract and 1 µL of 1 mM 3-oxo-C6-HSL was placed in the center of the plate. DMSO was used as a negative control, and piericidin A isolated from *Streptomyces xanthocidicus* KPP01532 was used as a positive control [67]. The plates were incubated overnight at 28 °C.

3.4. Isolation and Identification of the Bioactive Compound

3.4.1. Experimental Procedures

Column chromatography was performed by silica gel (200–300 mesh, Merck, Darmstadt, Germany), and prep-HPLC was performed by a Waters system consisting of a 515 pump and 2996 PDA detector (Milford, MA, USA). NMR spectra were recorded on a Varian 500-MHz NMR spectrometer (Palo Alto, CA, USA), and ESIMS spectra were recorded on an LCQ Fleet Ion Trap MS spectrometer (Thermo Scientific, Madison, WI, USA). UPLC was performed using Waters ACQUITY UPLC^{TM}.

3.4.2. Bioassay-Guided Isolation

The crude extract (826.2 mg) was fractionated on a column (3 × 30 cm) packed with a silica gel using elution with *n*-hexane-EtOAc gradient ratios (1:0 to 0:1) to provide seven fractions (F1–F7). Among the seven fractions, F5 exhibited 100% DPPH radical-scavenging activity at a concentration of 500 µg/mL and was separated into five peak fractions (F5-P1–F5-P5) by prep-HPLC (5–30% ACN, 8 mL/min). The first subfraction (F5-P1) exhibited 100% DPPH radical-scavenging activity at a concentration of 10 µg/mL and was further purified by silica gel column chromatography (1 × 20 cm) using elution with CHCl$_3$-MeOH in gradient ratios (50:1–20:1) to provide target Compound 1 (F5-P1-2, 61.9 mg).

Gentisyl alcohol (**1**) ^{1}H NMR (DMSO-d_6, 500 MHz) δ_H 8.58 (2H, br s, overlap, OH-2 and OH-5), 6.75 (1H, d, *J* = 3.0 Hz, H-6), 6.55 (1H, d, *J* = 8.5 Hz, H-3), 6.43 (1H, d, *J* = 8.5, 3.1 Hz, H-4), 4.92 (1H, br s, OH-7), 4.41 (2H, br s, H-7); ^{13}C NMR (DMSO-d_6, 125 MHz) δ_C 150.2 (C-5), 146.8 (C-2), 129.8 (C-1), 115.5 (C-3), 114.4 (C-6), 113.7 (C-4), 58.7 (C-7); ESIMS (*m/z*) 139.2 [M − H]$^-$, 185.1 [M + COOH]$^-$.

4. Conclusions

The multi-gene phylogenetic analysis revealed that 28 *Arthrinium* strains isolated from various marine environments consisted of four known species and 11 new species candidates. Most of them exhibited strong antioxidant activity, and some showed antifungal, quorum sensing inhibition, and tyrosinase inhibition activity. It was demonstrated that marine algicolous *Arthrinium* spp. can support the regulation of reactive oxygen species in symbiotic algae and protect against pathogens and bacterial biofilm formation. They can form a symbiotic relationship in a way that increases the viability and environmental competitiveness of symbiotic algae in exchange for nutrients such as mannitol, which is the main photosynthetic product of brown algae [68]. A strong antioxidant compound was isolated from a new species, *Arthrinium* sp. 10 KUC21332, and identified to be gentisyl alcohol (**1**), and the antioxidant compound of *A. saccharicola* KUC21221 was identical. These results suggest that many *Arthrinium* spp. remain unexploited in the marine environment, and marine *Arthrinium* spp. are a great source of bioactive compounds. Other bioactive compounds from marine *Arthrinium* spp., including antifungal compounds, tyrosinase inhibitors, and QSIs, will be isolated and identified in the near future.

Supplementary Materials: The following are available online at http://www.mdpi.com/1660-3397/16/12/508/s1: Figure S1: ^{1}H NMR spectrum of Compound 1, Figure S2: ^{13}C NMR spectrum of Compound 1, Figure S3: LC chromatogram and UV spectrum of Compound 1, Figure S4: MS spectra of Compound 1, Figure S5: UPLC peaks and UV spectrum of *A. saccharicola* KUC21221 extract, *Arthrinium* sp. 10 KUC21332 extract, and Compound **1**. The arrows indicate gentisyl alcohol, Table S1: GenBank accession numbers of the reference sequences used in the phylogenetic analysis in this study, Table S2: Marine algicolous *Arthrinium* spp. exhibiting antifungal and tyrosinase inhibition activity.

Author Contributions: Conceptualization, Y.M.H., J.-H.H., and J.-J.K.; methodology, M.Y.P., J.E.K., J.-H.H., and S.M.R.; validation, B.S.K., D.L. and J.-J.K.; formal analysis, Y.M.H. and M.Y.P.; investigation, Y.M.H., K.K., and S.L.K.; resources, Y.W.L., C.K., and J.-J.K.; writing, original preparation, Y.M.H., K.K., S.M.R., S.L.K., M.Y.P., and J.E.K.; writing, review and editing, B.S.K., D.L., and J.-J.K.; visualization, S.M.R. and S.L.K.; supervision, B.S.K., D.L., and J.-J.K. project administration, J.-J.K.; funding acquisition, J.-J.K.

Funding: This research was funded by National Research Foundation of Korea (NRF) funded by the Korean government (MSIT) grant number 2017R1A2B4002071. This research was also funded by the National Institute of Biological Resources (NIBR) under the Ministry of Environment, Republic of Korea, Grant Number 201701104.

Conflicts of Interest: The authors declare no conflict of interest.

References

1. Hong, J.-H.; Jang, S.; Heo, Y.M.; Min, M.; Lee, H.; Lee, Y.M.; Lee, H.; Kim,, J.-J. Investigation of marine-derived fungal diversity and their exploitable biological activities. *Mar. Drugs* **2015**, *13*, 4137–4155. [CrossRef] [PubMed]
2. Ellis, M.B. *Dematiaceous Hyphomycetes*; Commonwealth Mycological Institute: Kew, UK, 1971.
3. Sharma, R.; Kulkarni, G.; Sonawane, M.S.; Shouche, Y.S. A new endophytic species of *arthrinium* (apiosporaceae) from *Jatropha podagrica*. *Mycoscience* **2014**, *55*, 118–123. [CrossRef]
4. Crous, P.W.; Groenewald, J.Z. A phylogenetic re-evaluation of *arthrinium*. *IMA Fungus* **2013**, *4*, 133–154. [CrossRef] [PubMed]
5. Seifert, K.A.; Gams, W. The genera of hyphomycetes. *Persoonia* **2011**, *27*, 119–129. [CrossRef] [PubMed]
6. Ellis, M. *More Dematiaceous Hyphomycetes*; CABI Publishing: Kew, UK, 1976; p. 507.
7. Wang, M.; Tan, X.-M.; Liu, F.; Cai, L. Eight new *arthrinium* species from china. *MycoKeys* **2018**, 1–24. [CrossRef]
8. Blunt, J.W.; Copp, B.R.; Keyzers, R.A.; Munro, M.H.; Prinsep, M.R. Marine natural products. *Nat. Prod. Rep.* **2015**, *32*, 116–211. [CrossRef] [PubMed]
9. Bugni, T.S.; Ireland, C.M. Marine-derived fungi: A chemically and biologically diverse group of microorganisms. *Nat. Prod. Rep.* **2004**, *21*, 143–163. [CrossRef]
10. Molinski, T.F.; Dalisay, D.S.; Lievens, S.L.; Saludes, J.P. Drug development from marine natural products. *Nat. Rev. Drug Discov.* **2009**, *8*, 6969–6985. [CrossRef]
11. Mousa, W.K.; Raizada, M.N. The diversity of anti-microbial secondary metabolites produced by fungal endophytes: An interdisciplinary perspective. *Front. Microbiol.* **2013**, *4*, 65. [CrossRef]
12. Tsukada, M.; Fukai, M.; Miki, K.; Shiraishi, T.; Suzuki, T.; Nishio, K.; Sugita, T.; Ishino, M.; Kinoshita, K.; Takahashi, K.; et al. Chemical constituents of a marine fungus, *Arthrinium sacchari*. *J. Nat. Prod.* **2011**, *74*, 1645–1649. [CrossRef]
13. Ebada, S.S.; Schulz, B.; Wray, V.; Totzke, F.; Kubbutat, M.H.; Müller, W.E.; Hamacher, A.; Kassack, M.U.; Lin, W.; Proksch, P. Arthrinins a–d: Novel diterpenoids and further constituents from the sponge derived fungus *arthrinium* sp. *Bioorg. Med. Chem.* **2011**, *19*, 4644–4651. [CrossRef] [PubMed]
14. Elissawy, A.M.; Ebada, S.S.; Ashour, M.L.; Özkaya, F.C.; Ebrahim, W.; Singab, A.B.; Proksch, P. Spiroarthrinols a and b, two novel meroterpenoids isolated from the sponge-derived fungus *arthrinium* sp. *Phytochem. Lett.* **2017**, *20*, 246–251. [CrossRef]
15. Wei, M.-Y.; Xu, R.-F.; Du, S.-Y.; Wang, C.-Y.; Xu, T.-Y.; Shao, C.-L. A new griseofulvin derivative from the marine-derived *arthrinium* sp. Fungus and its biological activity. *Chem. Nat. Compd.* **2016**, *52*, 1011–1014. [CrossRef]
16. Bao, J.; Zhai, H.; Zhu, K.; Yu, J.-H.; Zhang, Y.; Wang, Y.; Jiang, C.-S.; Zhang, X.; Zhang, Y.; Zhang, H. Bioactive pyridone alkaloids from a deep-sea-derived fungus *arthrinium* sp. Ujnmf0008. *Mar. Drugs* **2018**, *16*, 174. [CrossRef]
17. Bao, J.; He, F.; Yu, J.-H.; Zhai, H.; Cheng, Z.-Q.; Jiang, C.-S.; Zhang, Y.; Zhang, Y.; Zhang, X.; Chen, G. New chromones from a marine-derived fungus, *arthrinium* sp., and their biological activity. *Molecules* **2018**, *23*, 1982. [CrossRef]

18. Wang, J.; Wang, Z.; Ju, Z.; Wan, J.; Liao, S.; Lin, X.; Zhang, T.; Zhou, X.; Chen, H.; Tu, Z. Cytotoxic cytochalasins from marine-derived fungus *Arthrinium arundinis*. *Planta Med.* **2015**, *81*, 160–166. [CrossRef]
19. Park, M.S.; Oh, S.-Y.; Lee, S.; Eimes, J.A.; Lim, Y.W. Fungal diversity and enzyme activity associated with sailfin sandfish egg masses in korea. *Fungal Ecol.* **2018**, *34*, 1–9. [CrossRef]
20. Park, J.Y.; Cho, J.K.; Son, M.H.; Kim, K.M.; Han, K.H.; Park, J.M. Artificial spawning behavior and development of eggs, larvae and juveniles of the red spotted grouper, epinephelus akaara in korea. *Dev. Reprod.* **2016**, *20*, 31–40. [CrossRef]
21. Miao, L.; Kwong, T.F.N.; Qian, P.-Y. Effect of culture conditions on mycelial growth, antibacterial activity, and metabolite profiles of the marine-derived fungus *arthrinium* c.F. Saccharicola. *Appl. Microbiol. Biotechnol.* **2006**, *72*, 1063–1073. [CrossRef] [PubMed]
22. Huang, W.-Y.; Cai, Y.-Z.; Xing, J.; Corke, H.; Sun, M. A potential antioxidant resource: Endophytic fungi from medicinal plants. *Econ. Bot.* **2007**, *61*, 14–30. [CrossRef]
23. Wang, H.-M.D.; Chen, C.-C.; Huynh, P.; Chang, J.-S. Exploring the potential of using algae in cosmetics. *Bioresour. Technol.* **2015**, *184*, 355–362. [CrossRef] [PubMed]
24. Machu, L.; Misurcova, L.; Vavra Ambrozova, J.; Orsavova, J.; Mlcek, J.; Sochor, J.; Jurikova, T. Phenolic content and antioxidant capacity in algal food products. *Molecules* **2015**, *20*, 1118–1133. [CrossRef] [PubMed]
25. Smith, H.; Doyle, S.; Murphy, R. Filamentous fungi as a source of natural antioxidants. *Food Chem.* **2015**, *185*, 389–397. [CrossRef] [PubMed]
26. Bartasiute, A.; Westerink, B.H.; Verpoorte, E.; Niederländer, H.A. Improving the in vivo predictability of an on-line hplc stable free radical decoloration assay for antioxidant activity in methanol-buffer medium. *Free Radic. Biol. Med.* **2007**, *42*, 413–423. [CrossRef] [PubMed]
27. Skriptsova, A.V. Fucoidans of brown algae: Biosynthesis, localization, and physiological role in thallus. *Russ. J. Mar. Biol.* **2015**, *41*, 145–156. [CrossRef]
28. Raghukumar, S. *Fungi in Coastal and Oceanic Marine Ecosystems*; Springer: New York, NY, USA, 2017.
29. Gomaa, M.; Hifney, A.F.; Fawzy, M.A.; Issa, A.A.; Abdel-Gawad, K.M. Biodegradation of *Palisada perforata* (rhodophyceae) and *sargassum* sp.(phaeophyceae) biomass by crude enzyme preparations from algicolous fungi. *J. Appl. Phycol.* **2015**, *27*, 2395–2404. [CrossRef]
30. Yap, P.S.X.; Yiap, B.C.; Ping, H.C.; Lim, S.H.E. Essential oils, a new horizon in combating bacterial antibiotic resistance. *Open Microbiol. J.* **2014**, *8*, 6–14. [CrossRef]
31. Chang, T.-S. An updated review of tyrosinase inhibitors. *Int. J. Mol. Sci.* **2009**, *10*, 2440–2475. [CrossRef]
32. Kang, H.S.; Choi, J.H.; Cho, W.K.; Park, J.C.; Choi, J.S. A sphingolipid and tyrosinase inhibitors from the fruiting body of *Phellinus linteus*. *Arch. Pharmacal. Res.* **2004**, *27*, 742–750. [CrossRef]
33. Li, X.; Kim, M.K.; Lee, U.; Kim, S.-K.; Kang, J.S.; Choi, H.D.; Son, B.W. Myrothenones a and b, cyclopentenone derivatives with tyrosinase inhibitory activity from the marine-derived fungus *myrothecium* sp. *Chem. Pharm. Bull.* **2005**, *53*, 453–455. [CrossRef]
34. Chen, J.S.; Wei, C.I.; Marshall, M.R. Inhibition-mechanism of kojic acid on polyphenol oxidase. *J. Agric. Food Chem.* **1991**, *39*, 1897–1901. [CrossRef]
35. Olson, M.E.; Ceri, H.; Morck, D.W.; Buret, A.G.; Read, R.R. Biofilm bacteria: Formation and comparative susceptibility to antibiotics. *Can. J. Vet. Res.* **2002**, *66*, 86–92. [PubMed]
36. Rasmussen, T.B.; Skindersoe, M.E.; Bjarnsholt, T.; Phipps, R.K.; Christensen, K.B.; Jensen, P.O.; Andersen, J.B.; Koch, B.; Larsen, T.O.; Hentzer, M.; et al. Identity and effects of quorum-sensing inhibitors produced by *Penicillium* species. *Microbiology* **2005**, *151*, 1325–1340. [CrossRef] [PubMed]
37. Borchardt, S.A.; Allain, E.J.; Michels, J.J.; Stearns, G.W.; Kelly, R.F.; McCoy, W.F. Reaction of acylated homoserine lactone bacterial signaling molecules with oxidized halogen antimicrobials. *Appl. Environ. Microbiol.* **2001**, *67*, 3174–3179. [CrossRef] [PubMed]
38. Kim, J.S.; Kim, Y.H.; Seo, Y.W.; Park, S. Quorum sensing inhibitors from the red alga, *Ahnfeltiopsis flabelliformis*. *Biotechnol. Bioprocess Eng.* **2007**, *12*, 308–311. [CrossRef]
39. Kalia, V.C. Quorum sensing inhibitors: An overview. *Biotechnol. Adv.* **2013**, *31*, 224–245. [CrossRef] [PubMed]
40. Baldry, M.; Nielsen, A.; Bojer, M.S.; Zhao, Y.; Friberg, C.; Ifrah, D.; Heede, N.G.; Larsen, T.O.; Frøkiær, H.; Frees, D.; et al. Norlichexanthone reduces virulence gene expression and biofilm formation in *Staphylococcus aureus*. *PLoS ONE* **2016**, *11*, e0168305. [CrossRef]
41. Skandamis, P.N.; Nychas, G.-J.E. Quorum sensing in the context of food microbiology. *Appl. Environ. Microbiol.* **2012**, *78*, 5473–5482. [CrossRef]

42. Schultz, M.P.; Bendick, J.A.; Holm, E.R.; Hertel, W.M. Economic impact of biofouling on a naval surface ship. *Biofouling* **2011**, *27*, 87–98. [CrossRef]

43. Chen, L.; Fang, Y.; Zhu, T.; Gu, Q.; Zhu, W. Gentisyl alcohol derivatives from the marine-derived fungus *Penicillium terrestre*. *J. Nat. Prod.* **2008**, *71*, 66–70. [CrossRef]

44. Nenkep, V.N.; Yun, K.; Li, Y.; Choi, H.D.; Kang, J.S.; Son, B.W. New production of haloquinones, bromochlorogentisylquinones a and b, by a halide salt from a marine isolate of the fungus *phoma herbarum*. *J. Antibiot.* **2010**, *63*, 199–201. [CrossRef] [PubMed]

45. Kim, H.J.; Kim, J.H.; Lee, C.H.; Kwon, H.J. Gentisyl alcohol, an antioxidant from microbial metabolite, induces angiogenesis in vitro. *J. Microbiol. Biotechnol.* **2006**, *16*, 475–479.

46. Kim, J.; Kim, D.; Kim, M.; Kwon, H.; Oh, T.; Lee, C. Gentisyl alcohol inhibits apoptosis by suppressing caspase activity induced by etoposide. *J. Microbiol. Biotechnol.* **2005**, *15*, 532–536.

47. Lernoux, M.; Schnekenburger, M.; Dicato, M.; Diederich, M. Anti-cancer effects of naturally derived compounds targeting histone deacetylase 6-related pathways. *Pharmacol. Res.* **2018**, *129*, 337–356. [CrossRef] [PubMed]

48. Zwick, V.; Allard, P.M.; Ory, L.; Simões-Pires, C.A.; Marcourt, L.; Gindro, K.; Wolfender, J.L.; Cuendet, M. Uhplc-ms-based hdac assay applied to bio-guided microfractionation of fungal extracts. *Phytochem. Anal.* **2017**, *28*, 93–100. [CrossRef]

49. Li, Y.; Li, X.; Son, B.-W. Antibacterial and radical scavenging epoxycyclohexenones and aromatic polyols from a marine isolate of the fungus *aspergillus*. *Nat. Prod. Sci.* **2005**, *11*, 136–138.

50. Gupta, S.; Kaul, S.; Singh, B.; Vishwakarma, R.A.; Dhar, M.K. Production of gentisyl alcohol from *Phoma herbarum* endophytic in *Curcuma longa* L. And its antagonistic activity towards leaf spot pathogen *Colletotrichum gloeosporioides*. *Appl. Biochem. Biotechnol.* **2016**, *180*, 1093–1109. [CrossRef] [PubMed]

51. Malak, L.G.; Ibrahim, M.A.; Bishay, D.W.; Abdel-baky, A.M.; Moharram, A.M.; Tekwani, B.; Cutler, S.J.; Ross, S.A. Antileishmanial metabolites from *Geosmithia langdonii*. *J. Nat. Prod.* **2014**, *77*, 1987–1991. [CrossRef]

52. Ali, T.; Inagaki, M.; Chai, H.-B.; Wieboldt, T.; Rapplye, C.; Rakotondraibe, L.H. Halogenated compounds from directed fermentation of *penicillium concentricum*, an endophytic fungus of the liverwort *Trichocolea tomentella*. *J. Nat. Prod.* **2017**, *80*, 1397–1403. [CrossRef]

53. Smetanina, O.; Kalinovskii, A.; Khudyakov, Y.V.; Moiseenko, O.; Pivkin, M.; Menzorova, N.; Sibirtsev, Y.T.; Kuznetsova, T. Metabolites of the marine fungus *Asperigillus varians* kmm 4630. *Chem. Nat. Compd.* **2005**, *41*, 243–244. [CrossRef]

54. Gardes, M.; Bruns, T.D. Its primers with enhanced specificity for basidiomycetes—Application to the identification of mycorrhizae and rusts. *Mol. Ecol.* **1993**, *2*, 113–118. [CrossRef] [PubMed]

55. Bellemain, E.; Carlsen, T.; Brochmann, C.; Coissac, E.; Taberlet, P.; Kauserud, H. Its as an environmental DNA barcode for fungi: An in silico approach reveals potential pcr biases. *BMC Microbiol* **2010**, *10*, 189. [CrossRef] [PubMed]

56. Vilgalys, R.; Hester, M. Rapid genetic identification and mapping of enzymatically amplified ribosomal DNA from several cryptococcus species. *J. Bacteriol.* **1990**, *172*, 4238–4246. [CrossRef] [PubMed]

57. O'Donnell, K.; Cigelnik, E. Two divergent intragenomic rdna its2 types within a monophyletic lineage of the fungus fusarium are nonorthologous. *Mol. Phylogenet. Evol.* **1997**, *7*, 103–116. [CrossRef] [PubMed]

58. Glass, N.L.; Donaldson, G.C. Development of primer sets designed for use with the pcr to amplify conserved genes from filamentous ascomycetes. *Appl. Environ. Microbiol.* **1995**, *61*, 1323–1330. [PubMed]

59. O'Donnell, K.; Kistler, H.C.; Cigelnik, E.; Ploetz, R.C. Multiple evolutionary origins of the fungus causing panama disease of banana: Concordant evidence from nuclear and mitochondrial gene genealogies. *Proc. Natl. Acad. Sci. USA* **1998**, *95*, 2044–2049. [CrossRef] [PubMed]

60. Carbone, I.; Kohn, L.M. A method for designing primer sets for speciation studies in filamentous ascomycetes. *Mycologia* **1999**, *91*, 553–556. [CrossRef]

61. Maddison, D.R.; Maddison, W.P. *Macclade 4: Analysis of Phylogeny and Character Evolution. Version 4.08 a*; Sinauer Associates: Sunderland, MA, USA, 2005.

62. Katoh, K.; Standley, D.M. Mafft multiple sequence alignment software version 7: Improvements in performance and usability. *Mol. Biol. Evol.* **2013**, *30*, 772–780. [CrossRef]

63. Nylander, J.A.A. *Mrmodeltest v2. Program Distributed by the Author*; Evolutionary Biology Centre: Uppsala, Sweden, 2004.

64. Ronquist, F.; Huelsenbeck, J.P. Mrbayes 3: Bayesian phylogenetic inference under mixed models. *Bioinformatics* **2003**, *19*, 1572–1574. [CrossRef]
65. Kim, J.D.; Han, J.W.; Hwang, I.C.; Lee, D.; Kim, B.S. Identification and biocontrol efficacy of *Streptomyces miharaensis* producing filipin iii against *Fusarium wilt*. *J. Basic Microbiol.* **2012**, *52*, 150–159. [CrossRef]
66. Lai, H.Y.; Lim, Y.Y.; Tan, S.P. Antioxidative, tyrosinase inhibiting and antibacterial activities of leaf extracts from medicinal ferns. *Biosci. Biotechnol. Biochem.* **2009**, *73*, 1362–1366. [CrossRef] [PubMed]
67. Kang, J.E.; Han, J.W.; Jeon, B.J.; Kim, B.S. Efficacies of quorum sensing inhibitors, piericidin a and glucopiericidin a, produced by *Streptomyces xanthocidicus* kpp01532 for the control of potato soft rot caused by *Erwinia carotovora* subsp. *atroseptica*. *Microbiol. Res.* **2016**, *184*, 32–41. [CrossRef] [PubMed]
68. Yamaguchi, T.; Ikawa, T.; Nisizawa, K. Pathway of mannitol formation during photosynthesis in brown algae. *PCPhy* **1969**, *10*, 425–440.

 *marine drugs*

Article

Fungal Diversity in Intertidal Mudflats and Abandoned Solar Salterns as a Source for Biological Resources

Young Mok Heo [1,†], Hanbyul Lee [1,†], Kyeongwon Kim [1], Sun Lul Kwon [1], Min Young Park [2], Ji Eun Kang [2], Gyu-Hyeok Kim [1], Beom Seok Kim [3] and Jae-Jin Kim [1,*]

1   Division of Environmental Science & Ecological Engineering, College of Life Sciences & Biotechnology, Korea University, Seoul 02841, Korea; hym011@korea.ac.kr (Y.M.H.); hblee95@korea.ac.kr (H.L.); rudndjs@korea.ac.kr (K.K.); sun-lul@korea.ac.kr (S.L.K.); lovewood@korea.ac.kr (G.-H.K.)
2   Department of Biosystems & Biotechnology, College of Life Sciences and Biotechnology, Korea University, Seoul 02841, Korea; min0@korea.ac.kr (M.Y.P.); heyyo9725@korea.ac.kr (J.E.K.)
3   Division of Biotechnology, College of Life Sciences & Biotechnology, Korea University, Seoul 02841, Korea; biskim@korea.ac.kr
*   Correspondence: jae-jinkim@korea.ac.kr; Tel.: +82-2-3290-3049; Fax: +82-2-3290-9753
†   These authors contributed equally to the article.

Received: 26 September 2019; Accepted: 22 October 2019; Published: 23 October 2019

Abstract: Intertidal zones are unique environments that are known to be ecological hot spots. In this study, sediments were collected from mudflats and decommissioned salterns on three islands in the Yellow Sea of South Korea. The diversity analysis targeted both isolates and unculturable fungi via Illumina sequencing, and the natural recovery of the abandoned salterns was assessed. The phylogeny and bioactivities of the fungal isolates were investigated. The community analysis showed that the abandoned saltern in Yongyudo has not recovered to a mudflat, while the other salterns have almost recovered. The results suggested that a period of more than 35 years may be required to return abandoned salterns to mudflats via natural restoration. Gigasporales sp. and *Umbelopsis* sp. were selected as the indicators of mudflats. Among the 53 isolates, 18 appeared to be candidate novel species, and 28 exhibited bioactivity. *Phoma* sp., *Cladosporium sphaerospermum*, *Penicillium* sp. and *Pseudeurotium bakeri*, and *Aspergillus urmiensis* showed antioxidant, tyrosinase inhibition, antifungal, and quorum-sensing inhibition activities, respectively, which has not been reported previously. This study provides reliable fungal diversity information for mudflats and abandoned salterns and shows that they are highly valuable for bioprospecting not only for novel microorganisms but also for novel bioactive compounds.

Keywords: fungal community; marine fungi; phylogenetic analysis; saltwork; tidal flat

1. Introduction

As interest in marine living resources worldwide has increased, research on marine fungi has progressed considerably over the past two decades, including the discovery of new species and novel natural compounds [1]. It is still insufficient compared to the research on terrestrial fungi, but the study of marine fungi has been extended to intertidal zones such as mangrove forests and coastal wetlands [2,3]. In addition, there are other unique intertidal environments such as mudflats and abandoned salterns.

There are different types of tidal flats, and mudflats are differentiated from sandy tidal flats. Mudflats are a kind of coastal salt marsh made of clay deposited by waves and rivers and are a unique environment as they are exposed to the atmosphere twice a day depending on the tide. In the Yellow

Sea of South Korea, mudflats are distributed widely along the coastline. The Ministry of Oceans and Fisheries (MOF, South Korean government) reported that the total area of mudflats in 2013 was 2487.2 km^2, which is approximately 2.5% of the national territory [4]. Although the area has decreased by 22.4% since 1987 due to the mudflat reclamation project for the expansion of agriculture land, recent efforts have been made to preserve and restore mudflat areas. For example, the MOF expanded the protected mudflat area from 79.62 km^2 to approximately 1265.46 km^2 in 2018 [5]. The coast of the Yellow Sea has topographical conditions that are suitable for solar salt production, as the coastal slope is very low. A large number of solar salterns have been constructed and operated on the coast throughout history. A broad area of the intertidal mudflat had even been reclaimed for this purpose. However, in the 1980s, people were informed that mudflats have great ecological and economic value, so citizens and environmental scientists insisted on restoring the mudflats. Mudflats are an ecological hot spot that performs numerous ecological functions, such as providing habitat for shellfish, water birds, and migratory birds. According to the Korea Marine Environment Management Corporation (KOEM, South Korean government), the economic value of the Korean mudflats was estimated at approximately 15 billion dollars per year in 2017, including value from food production, waste treatment, recreation, habitat/refugia, disturbance regulation, and conservation [6]. In addition, people were concerned about consuming solar salt, as global concerns about marine pollution and its negative effects on human health were raised. In this context, most of the salterns were closed because of the government's policy as well as the decrease in market demand, and the embankments of the closed salterns were withdrawn to allow the natural recovery of the mudflats [5]. As a result, these abandoned salterns have become a unique environment where the tides have returned to the salterns installed in the dry and windy areas of the intertidal zone.

Because of strong selective pressure, environments that are unique or disadvantageous for survival are likely to have novel microorganisms as well as novel secondary metabolites [7,8]. Therefore, microorganisms living in these unusual environments, especially fungi, which have been considered a treasure trove of bioactive secondary metabolites, are highly valuable to investigate. Many studies have been reported on saltern extremophiles and microorganisms in the mangrove forest, another intertidal environment [9–12]. Until now, however, studies of the diversity, physiology, and bioactivity of fungi in intertidal mudflats and abandoned salterns have rarely been reported in the overall field. To the best of our knowledge, there are only two studies on the fungal diversity of these environments: one on the diversity of *Aspergillus* in a mudflat and one on the diversity of mycorrhizal fungi in an abandoned saltern [13,14]. Therefore, it is necessary to investigate the microbial communities of these environments. Through community analysis, a recovery assessment of the decommissioned salterns is also possible; there has been no study on the recovery of the microbial community in this environment to date [15]. Even in other similar sites around the world, few studies on macrofauna/flora have been reported, and there has been no study of the microorganisms, which are the cornerstone of the ecosystem. In fact, it has been proven by many researchers in recent decades that microbial community analysis is a useful tool for ecosystem assessment. Most such studies have used bacterial communities because of the large bacterial DNA database that is available. However, a large amount of fungal DNA data has now been stored as well. It is also necessary to evaluate these environments as biological resources. Studies on the fungi isolated from mudflats and their bioactive secondary metabolites are also rare. To date, *Chaetomium cristatum*, *Fusarium oxysporum*, *Aspergillus niger*, *Thielavia hyalocarpa*, *Thielavia* sp., and *Paecilomyces formosus* have been reported and are known to produce cristazine (cytotoxic and antioxidant), oxysporizoline (antibacterial), 6,9-dibromoflavasperone (antioxidant), 1-*O*-(α-D-mannopyranosyl)geraniol (a biotransformation product of geraniol), thielaviazoline (antibacterial and antioxidant), and formoxazine (antibacterial and antioxidant), respectively [16–21]. The lack of studies is likely primarily due to the limited global distribution of mudflats and salterns. Another possible reason is that it is difficult to isolate a large number of fungi due to the extreme environmental conditions; the environment dramatically changes between aerobic and anaerobic conditions twice a day, and dry stress occurs at the ebb of the tide due to the weather conditions described above.

The aims of this study were 1) to analyze the fungal community and diversity in the intertidal mudflats and the abandoned solar salterns of the Yellow Sea in South Korea; 2) to assess the natural recovery of the abandoned salterns to mudflats and determine the indicator taxa; and 3) to evaluate these environments for the bioprospecting of novel species and bioactive compounds.

2. Results and Discussion

2.1. The Fungal Community in Intertidal Mudflats and Abandoned Salterns

Intertidal sediments were sampled from mudflats and abandoned salterns in Yongyudo, Gopado, and Yubudo located on the coast of the Yellow Sea in South Korea (Figure 1).

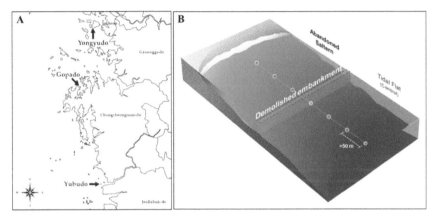

Figure 1. (**A**) Locations of the three sampling sites and an illustration of abandoned salterns and mudflats in the west coast of South Korea. (**B**) Illustration of abandoned salterns and tidal flats in the west coast of South Korea. The circles indicate the sampling points.

2.1.1. Fungal Diversity and Recovery Assessment

The fungal communities and diversity in these intertidal environments were investigated by high-throughput sequencing of the sediment samples. Through the community analysis, the natural recovery of the decommissioned salterns was also assessed.

At the phylum level of the fungal community composition, Ascomycota and Basidiomycota, the two major phyla of the kingdom Fungi, were the dominant phyla (Figure 2A). Meanwhile, the community composition of the abandoned saltern in Yongyudo was clearly different. It was highly dominated by Entorrhizomycota, while Chytridiomycota and Mortierellomycota dominated the other abandoned salterns and mudflats. Considering that the phylum Entorrhizomycota is composed of plant pathogenic fungi, the difference may be due to the introduction of their host plants at the initial stage of ecological succession after the saltern was abandoned. Meanwhile, a large number of fungi could not be identified, indicating a lack of DNA-based phylogenetic information on fungi in these intertidal environments. As the community compositions of the other salterns and mudflats were similar to each other except for that of the saltern in Yongyudo, a nonmetric multidimensional scaling (NMDS) plot was constructed using the database of fungal operational taxonomical units (OTUs) and their abundance to determine the distance between the fungal communities. According to the NMDS results, it appeared that the distance between the abandoned saltern and the mudflats in Yongyudo was much longer than the distance between the two environments on the other islands (Figure 2B). This indicates that the abandoned salterns in Yongyudo that were abandoned less than a year ago were the least recovered to mudflat. Since the salterns of Yubudo and Gopado were abandoned 20 and 35 years ago, respectively, it was expected that natural recovery would have occurred. This was supported by the short distance between the communities of the two salterns and those of nearby mudflats on the NMDS plot. Thus, we clustered the samples by the expected recovery status, consisting of recovered saltern (RS), nonrecovered saltern (NS), and mudflat. In fact, the NMDS plot showed that the fungal communities of the two environment types in Yubudo and Gopado were almost similar but were not completely clustered, especially in Gopado. This indicated that the fungal communities of the abandoned salterns were not fully recovered even after 35 years. Similarly, Bernhard and his colleague reported that salt marshes that were impounded and subsequently restored took more than 30 years to recover and become similar to the neighboring tidal flats [22]. They suggested that the regular inflow of tidal water increases the stability of the ecosystem. The abandoned salterns require a great deal of time to return to their original state due to the disturbance of human activity blocking the tidal water for a long time, resulting in a significant decrease in stability of microbial communities. Thus, with natural restoration alone, an artificially isolated environment requires an extremely long time to construct a microbial ecosystem similar to that of the adjacent environment, and it may require more than 35 years in the case of the abandoned salterns.

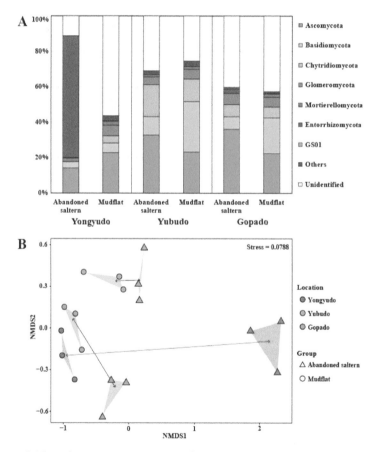

Figure 2. (**A**) Fungal community composition of intertidal mudflats and abandoned salterns in Yongyudo, Yubudo, and Gopado at the phylum level. Mean values of the abundance are used. (**B**) Nonmetric multidimensional scaling plot of the fungal community in intertidal mudflats and abandoned salterns in Yongyudo, Yubudo, and Gopado.

To verify the differences between the fungal communities, Shannon-Wiener and Gini-Simpson indices of each community were calculated. These two indices have been commonly used to calculate α-diversity, and the value of Shannon-Wiener and Gini-Simpson indices are more affected by species richness and evenness, respectively. The α-diversity indices of the NS were significantly lower in both the indices compared to those of the other environments, as expected (Table 1). This indicates that artificially blocking tidal inflow and conducting solar salt production reduce the species richness and evenness of the microbial community in the saltern, and the reduced diversity does not recover in a short period of time.

Table 1. The α-diversity indices of the fungal community in abandoned salterns and intertidal mudflats.

Sample	Shannon-Wiener Index (H')	Gini-Simpson Index (D^S, 1-λ)
Gopado		
Abandoned saltern (**RS**)	4.177	0.928
Mudflat	3.607	0.885
Yubudo		
Abandoned saltern (**RS**)	4.433	0.954
Mudflat	3.901	0.806
Yongyudo		
Abandoned saltern (**NS**)	1.471*	0.477*
Mudflat	3.497	0.859

Mean values are presented. NS, non-recovered saltern; RS, recovered saltern. * statistical significance of $p < 0.05$.

We tested the significance of the difference by the clustered group using the Kruskal-Wallis rank sum test. There was no significant difference between the α-diversity indices of RSs and mudflats, while those of the NS were significantly lower than those of the others. The abandoned saltern in Yongyudo was an environment continuously interrupted by humans until recently. Thus, the ecosystem of the NS could be easily dominated by a small number of species that could survive or adapt under this harsh condition, resulting in significantly lower microbial diversity. In fact, the diversity indices of RSs and mudflats were high when compared to those of fungal communities in other environments, such as livestock manure, compost, and soil [23–26]. This implies that diverse fungi exist in these unique environments, thus making them valuable for bioprospecting.

2.1.2. Determination of Indicator Taxa

We tried to determine the major differences among the fungal communities. Various analytical methods were used to find indicator taxa that are responsible for the differences, including linear discriminant analysis effect size (LEfSe), indicator species analysis (ISA), RandomForest, and mvabund.

First, we visualized the result of the Kruskal-Wallis rank sum test with a LEfSe cladogram (Figure 3 and Figure S1). It was obvious that the abundance of Ascomycota differs significantly between NS and the other environmental groups, tending to be lower in NS (Figure S2). RS and mudflat both tended to be abundant in Chytridiomycota and Basidiomycota, but many of the taxa overlapped each other. This finding supports that there is no significant difference between RSs and mudflat, and the RSs were in the process of recovery to mudflats.

For the next step, we investigated the taxa specific to each group via ISA. ISA results in an "IndVal score" calculated by multiplying the relative abundance by the relative frequency of a taxon in each group. Microbial community data are known to be zero-rich and to have many rare species. In this case, OTUs with low mean abundance tend to have low mean variance; this can lead to statistical misinterpretations if some rare species are highly specific to a certain environment. Therefore, the results with low frequency have low fidelity, and taxa with frequencies less than 5 were screened out. As a result, several lineages of taxa were selected to be specific to each environment (Table 2).

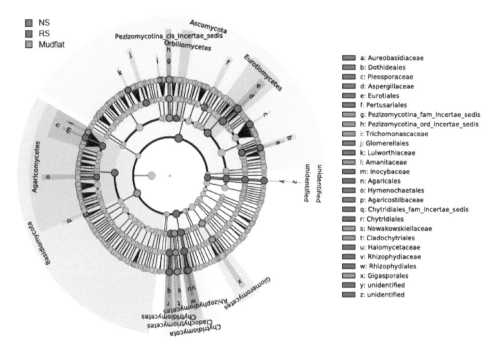

Figure 3. Linear discriminant analysis effect size (LEfSe) cladogram of fungal composition in abandoned salterns and mudflats (logarithmic linear discriminant analysis (LDA) score > 2, *p* < 0.05). Red, green, and blue nodes/shades indicate taxa that are significantly higher in relative abundance. The diameter of each node is proportional to the taxon's abundance. NS and RS mean non-recovered saltern and recovered saltern, respectively.

Table 2. Indicator taxa results from the indicator species analysis (IndVal score > 0.9, *p* < 0.05, and Frequency ≥ 5).

Taxa	Group	IndVal Score	*p* Value	Frequency
Ascomycota				
Stemphylium	NS	1.0000	0.005	8
Lecanoromycetes	NS	0.9012	0.007	17
Pertusariales	NS	0.9266	0.005	16
Basidiomycota				
Chionosphaeraceae	RS	0.9137	0.005	15
Ballistosporomyces sasicola	RS	0.9137	0.006	13
Tremellales	NS	0.9238	0.005	12
Tremellaceae	NS	0.9362	0.004	12
Tremella	NS	0.9359	0.005	9
Chytridiomycota				
Zygorhizidium planktonicum	RS	0.9322	0.003	8
Entorrhizomycota	NS	0.9995	0.004	16
Entorrhizomycetes	NS	0.9995	0.003	16
Mucoromycota				
Umbelopsidomycetes	Mudflat	0.9553	0.001	5
Umbelopsis sp.	Mudflat	0.9632	0.001	5

NS, non-recovered saltern; RS, recovered saltern.

RandomForest analysis was carried out to determine whether there is a hierarchy-based relationship in the fungal community that can distinguish the sample groups. The results are shown as the mean decrease in accuracy and Gini impurity, which indicates the importance of the taxa in the generated decision tree. At the class to species level, two lineages of taxa, Gigasporales sp. and *Umbelopsis* sp., were repeatedly ranked at the top (Figure 4).

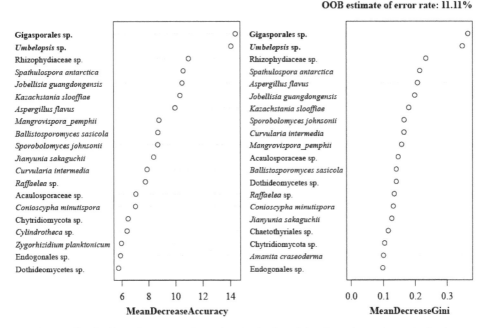

Figure 4. RandomForest importance plot analyzed using fungal abundance data at the species level.

Finally, mvabund was used to determine which taxon abundances differed significantly by group. Recently, it was proved that model-based analysis of multivariate abundance data combined with negative binomial regression is excellent for detecting multivariate effects that otherwise would lead to statistical misinterpretation and is powerful when selecting indicator species using multivariate abundance data [27]. Several lineages of taxa, including the order Gigasporales and the family Umbelopsidaceae, appeared to differ significantly among the groups (Table 3).

Various analyses were conducted to determine indicator taxa, but the results of these analyses were not identical. Therefore, the indicator species supported by multiple analyses were selected. Gigasporales sp. was supported by every analysis except for ISA, and *Umbelopsis* sp. was selected in every analysis except for LEfSe. Thus, these two fungal lineages can be regarded as definite indicator taxa for mudflats. In fact, the order Gigasporales belongs to the Glomeromycetes (Glomeromycota), the arbuscular mycorrhizal fungi. It was speculated that the abundance of Gigasporales sp. increases as it forms a symbiotic network with the indigenous halophytes of mudflats and as the abandoned saltern is restored to a mudflat [28]. Meanwhile, Entorrhizomycetes was selected by both ISA and mvabund as well as by a simple visual assessment of the fungal community composition (Figure 2A). However, it was difficult to determine the indicator taxa for the NS because only one sampling site, the abandoned saltern in Yongyudo, represented the NS environment. In other words, this plant pathogenic fungus may be dominant there because of the existence of its unique host plants. Therefore, it can be regarded as an indicator taxon of the NS in Yongyudo but not of all the NSs in the Yellow Sea.

Table 3. Indicator taxon results from the mvabund analysis ($p < 0.05$).

Taxa	Dev.	p Value
Ascomycota		
Leptosporella	26.15	0.035
Basidiomycota		
Agaricostilbomycetes	23.23	0.011
Chytridiomycota	21.43	0.011
Rhizophydiales	33.15	0.008
Rhizophydiaceae	31.84	0.013
Chytridiomycetes	23.13	0.013
Entorrhizomycota	34.62	0.004
Entorrhizomycetes	34.62	0.003
Entorrhizaceae	30.59	0.015
Entorrhiza cypericola	27.29	0.001
Glomeromycota		
Gigasporales	36.90	0.005
Mucoromycota		
Umbelopsidomycetes	22.20	0.018
Umbelopsidaceae	23.36	0.047
Rozellomycota	16.23	0.033

2.1.3. Diversity of Culturable Fungi

The community analysis implied that these intertidal environments are valuable for bioprospecting. Thus, the culturable fungi were isolated from the sediments to evaluate their diversity and to compare the results of metagenome analysis.

A total of 53 fungal strains was isolated from the sediments of the intertidal mudflats and the abandoned salterns (Table 4). It was obvious that the number of fungal isolates was much higher in Yongyudo than in the other two regions. It was suspected that this could be due to the sediment properties, such as the texture and grain size, since there was no significant difference in the environmental factors among sampling locations (Figure S1). The genera *Talaromyces* and *Trichoderma* were mostly isolated from the mudflat samples, and most *Penicillium* spp. were found in the NS. Meanwhile, the diversity and the number of fungal isolates in the NS were much higher than those of the other salterns, which was completely opposite to the result of the metagenome analysis, where the diversity of the NS was the lowest (Table 1, Table 4 and Table S1). In addition, all of the isolates except for *Phanerochaete chrysosporium* KUC10791 belong to Ascomycota. These results support that the actual diversity of environmental microorganisms must be investigated by eDNA-based metabarcoding approaches, unless all possible culture conditions and isolation methods are employed [29].

All 53 strains were grouped into 43 groups by morphological analysis and ITS sequences. The best-fit model of ITS sequences is a general time reversible (GTR) + proportion of invariable sites (I) + gamma distribution (G) by MrModeltest and contains 121 taxa and 844 nucleotide characters. In the phylogenetic analysis, 53 strains were classified into 2 phyla, 5 classes, 11 orders, 20 families, 22 genera, and 43 species, based on current taxonomic concepts. The dominant genera of the ITS tree were *Penicillium* (number of strains = 9), followed by *Talaromyces* (number of strains = 7), *Aspergillus* (number of strains = 6), *Trichoderma* (number of strains = 6), and *Cladosporium* (number of strains = 2) (Figure 5). *Talaromyces* and *Trichoderma* were the most complex clades and had low resolution. Thus, to precisely perform phylogenetic analysis of *Trichoderma* and *Talaromyces*, EF1-α for *Trichoderma* species and *benA* for *Talaromyces* species were amplified, and phylogenetic trees were constructed (Figure S3 and S4). The best-fit model of EF1-α for *Trichoderma* is Hasegawa-Kishino-Yano (HKY) + I + G and contains 41 taxa and 794 nucleotide characters. Multiple sequence alignments of two loci (ITS & *benA*) were analyzed for *Talaromyces*. Both loci were assigned GTR + I + G as the best fit model, and concatenated datasets contained 64 taxa and 1,155 nucleotide characters (ITS, nchar = 598; *benA*, nchar = 557). Through the EF1-α phylogenetic analysis of *Trichoderma*, KUC21406, KUC21404, KUC21401, KUC21411, and KUC21394 are confidently classified as *T. afroharzianum* and *T. harzianum* in the EF1-α tree. *Talaromyces* sp. 2 KUC21413 was closely related to *T. viridulus* in the ITS tree. In the concatenated tree (ITS & *benA*), however, the position changed to be near *T. galapagensis* with low posterior probability. Therefore, *Talaromyces* sp. 2 KUC21413 was assigned as a novel species candidate. *Talaromyces* sp. 1 KUC21276 appeared to be closely related to *T. angelicus* (99.81% sequence similarity in ITS; 93.22% in benA). *Talaromyces* sp. 3 KUC21415 and *Talaromyces* sp. 3 KUC21421 appeared to be closely related to *T. helices* (99.81% sequence similarity in ITS; 96.7% in *benA*). *Talaromyces* sp. 4 KUC21408 appeared to be closely related to *T. boninensis* (97.97% sequence similarity in ITS; 93.72% in *benA*). Thus, they are suggested to be candidate novel species.

Table 4. General information on the 53 fungal isolates from sediments of abandoned salterns and intertidal mudflats.

Fungal Name	KUC ID	Isolation Source		GenBank Accession Number		
		Location	Environmental Type	ITS	*benA*	EF1-α
Acremonium sp.	21384			MN518379	N.D.	N.D.
Aspergillus urmiensis	21379			MN518380	N.D.	N.D.
A. urmiensis	21392			MN518381	N.D.	N.D.
A. urmiensis	21396			MN518382	N.D.	N.D.
Cladosporium sphaerospermum	21388			MN518383	N.D.	N.D.
Emericellopsis microspora	21381			MN518384	N.D.	N.D.
Gibellulopsis nigrescens	21385			MN518385	N.D.	N.D.
Humicola alopallonella	21393			MN518386	N.D.	N.D.
Lulwoana sp.	21398			MN518387	N.D.	N.D.
Nectriaceae sp.	21383			MN518388	N.D.	N.D.
Neocamarosporium betae	21428			MN518389	N.D.	N.D.
Penicillium chrysogenum	21395		Abandoned saltern (NS)	MN518390	N.D.	N.D.
Penicillium citrinum	21390			MN518391	N.D.	N.D.
Penicillium sumatrense	21382			MN518392	N.D.	N.D.
Penicillium sp. 1	21380			MN518393	N.D.	N.D.
Penicillium sp. 1	21386			MN518394	N.D.	N.D.
Penicillium sp. 1	21387			MN518395	N.D.	N.D.
Penicillium sp. 1	21389			MN518396	N.D.	N.D.
Penicillium sp. 1	21399			MN518397	N.D.	N.D.
Phanerochaete chrysosporium	10791	Yongyudo		MN518398	N.D.	N.D.
Simplicillium aogashimaense	21397			MN518399	N.D.	N.D.
Sporormiaceae sp.	21391			MN518400	N.D.	N.D.
Trichoderma harzianum	21394			MN518401	N.D.	MN580170
Acremonium persicinum	21416			MN518402	N.D.	N.D.
Amorphotheca resinae	21414			MN518403	N.D.	N.D.
Aspergillus floccosus	21405			MN518404	N.D.	N.D.
Helotiaceae sp.	21409			MN518405	N.D.	N.D.
Penicillium sp. 2	21400			MN518406	N.D.	N.D.
Pseudogymnoascus pannorum	21417			MN518407	N.D.	N.D.
Sporothrix mexicana	21410			MN518408	N.D.	N.D.
Talaromyces liani	21412			MN518409	MN531288	N.D.
Talaromyces stipitatus	21402		Mudflat	MN518410	MN531294	N.D.
Talaromyces sp. 2	21413			MN518411	MN531292	N.D.
Talaromyces sp. 3	21415			MN518412	MN531289	N.D.
Talaromyces sp. 4	21408			MN518413	MN531293	N.D.
Trichoderma koningiopsis	21403			MN518414	N.D.	MN580171
Trichoderma afroharzianum	21401			MN518415	N.D.	MN580166
T. afroharzianum	21404			MN518416	N.D.	MN580167
T. afroharzianum	20406			MN518417	N.D.	MN580168
Trichoderma harzianum	21411			MN518418	N.D.	MN580169
Westerdykella capitulum	21407			MN518419	N.D.	N.D.
Cladosporium sp.	21418		Abandoned saltern (RS)	MN518420	N.D.	N.D.
Leptosphaeria sp.	21419	Gopado		MN518421	N.D.	N.D.
Hypocreales sp.	21420		Mudflat	MN518422	N.D.	N.D.
Talaromyces sp. 3	21421			MN518423	MN531290	N.D.
Aspergillaceae sp.	21424			MN518424	N.D.	N.D.
Aspergillus japonicus	21425			MN518425	N.D.	N.D.
Aspergillus tritici	21427			MN518426	N.D.	N.D.
Paracamarosporium sp.	21423	Yubudo	Abandoned saltern (RS)	MN518427	N.D.	N.D.
Phoma sp.	21426			MN518428	N.D.	N.D.
Pseudeurotium bakeri	21422			MN518429	N.D.	N.D.
Talaromyces sp. 1	21276			MN518430	MN531291	N.D.
Monascus sp.	21277		Mudflat	MN518431	N.D.	N.D.

KUC ID, Korea University Culture collection ID; ITS, internal transcribed spacer; *benA*, β-tubulin; EF1-α, translation elongation factor 1-α; NS, non-recovered saltern; RS, recovered saltern; N.D., no data.

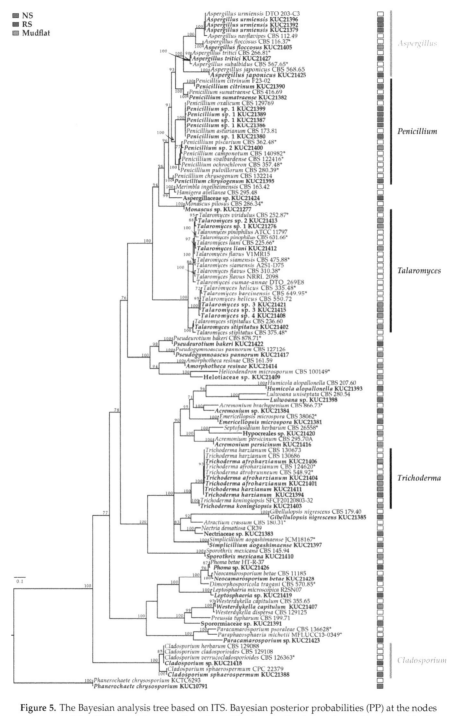

Figure 5. The Bayesian analysis tree based on ITS. Bayesian posterior probabilities (PP) at the nodes are presented if > 75. Type strains are indicated as *. The fungi isolated from this study are in bold. The scale bar indicates the number of nucleotide substitutions per position. NS and RS mean non-recovered saltern and recovered saltern, respectively.

The dominant taxa of the intertidal sediments were *Penicillium* species (number of strains = 8) and *Talaromyces* species (number of strains = 6). A number of fungal isolates belonged to Eurotiomycetes, followed by Sordariomycetes. A total of 18 candidate novel fungal species was isolated from this study. This result demonstrates that many novel fungal candidates remain unexploited in abandoned salterns or intertidal mudflat sediments.

2.2. Biological Activities of Fungi from Intertidal Mudflats and Abandoned Salterns

To evaluate the potential value of the intertidal environments for bioprospecting, a variety of biological activities of the fungal extracts were investigated.

2.2.1. Antioxidant Activity

It is well known that reactive oxygen species (ROS) and excessive free radicals cause oxidative chain reactions that damage biomolecules such as proteins and DNA and are responsible for a wide variety of diseases and aging [30,31]. Many antioxidants scavenging free radicals have been discovered, and fungi are one of the major sources of these compounds. However, only four antioxidants were discovered from mudflat-derived fungi: cristazine, 6,9-dibromoflavasperone, thielaviazoline, and formoxazine from *Chaetomium cristatum*, *Aspergillus niger*, *Thielavia* sp., and *Paecilomyces formosus*, respectively [16,18,20,21]. In this study, the extracts of 68 fungi isolated from intertidal mudflats and abandoned salterns were screened for their antioxidant capacity. Among the fungal extracts, three *Aspergillus* spp. (*A. floccosus* KUC21405, *A. japonicas* KUC21425, and *A. urmiensis* KUC21396), four *Penicillium* spp. (*P. chrysogenum* KUC21395 and *Penicillium* sp. 1 KUC21386, KUC21387, and KUC21389), *Phoma* sp. KUC21426, and *Talaromyces liani* KUC21421 exhibited high radical-scavenging activity in both assays using 2.2′-azino-bis-3-ethylbenzothiazoline-6-sulfonic acid (ABTS) radicals and 2,2-diphenyl-1-picrylhydrazyl (DPPH) radicals as substrates, respectively (Table 5).

Several antioxidants have been reported from the genus *Aspergillus*, for example, 3,3″-dihydroxyterphenyllin, 3-hydroxyterphenyllin, candidusin B, and dihydroxymethyl pyranone from *A. candidus* [32,33]; 1,8-dihydroxynaphthalene melanin from *A. bridgeri* [34]; methyl 4-(3,4-dihydroxybenz amido)butanoate, 5-O-methylsulochine, methyl 2-(2,6-dimethoxy-4-methylbenzoyl)-3,5-dihydroxybenz oate, methyl-2-(2,6-dihydroxyl-4-methylbenzoyl)-3-hydroxy-5-methoxybenzoate, physcion, 4-(3,4-dihy droxybenzamido)butanoic acid, and (E)-N-(2-hydroxy-2-(4-hydroxyphenyl)ethyl)-3-(3-hydroxy-4-methoxyphenyl)acrylamide from *A. wentii* [35]. However, there is no report of antioxidants from *A. floccosus*, *A. japonicus*, or *A. urmiensis* that showed the strong radical-scavenging activity seen in our results. The genus *Penicillium* has been reported to produce antioxidants, for example, atrovenetin from *P. atrovenetum*, *P. herquei*, and *P. simplicissimum* (previously known as *P. paraherquei*) [36–38]; 2,3,4-trimethyl-5,7-dihydroxy-2,3-dihydrobenzofuran and gentisic acid from *P. citrinum* [39]; and 2,3-dihydroxy benzoic acid from *P. roquefortii* [40]. An ethanol extract of *P. chrysogenum* showing moderate radical-scavenging activity in our results showed high activity in a previous study [41]. Because bisvertinolone has been reported in this species and many bisorbicillinoids are known to be antioxidants, the antioxidant of this species is likely to be bisvertinolone [42]. In the case of the genus *Phoma*, there is only one report of antioxidant compounds, for example, bromochlorogentisylquinones A and B, chlorogentisyl alcohol, and gentisyl alcohol from *P. herbarum* [43]. Some antioxidants have been reported from the genus *Talaromyces*, for example, 8-hydroxyconiothyrinone B, 8,11-dihydroxyconiothyrinone B, 4R,8-dihydroxyconiothyrinone B, 4S,8-dihydroxyconiothyrinone B, and 4S,8-dihydroxy-10-O-methyldendryol E from *T. islandicus* [44]; pentalsamonin from *T. purpureogenus* [45]; N-(4-hydroxy-2-methoxyphenyl) acetamide from *T. funiculosus* [46]; and 6-methylbiphenyl-3,3′,4,5′-tetraol and desmethylaltenusin [47].

Table 5. Fungal isolates from sediments of abandoned salterns and intertidal mudflats exhibiting radical-scavenging, tyrosinase inhibitory, and antifungal activity.

FUNGAL NAME	KUC ID	Antioxidant Activity (IC$_{50}$, µg/mL)		Tyrosinase Inhibitory Activity (IC$_{50}$, µg/mL)	Antifungal Activity (MIC, µg/mL)	
		ABTS	DPPH		Asteromyces cruciatus	Lindra thalassiae
Acremonium persicinum	21416	> 50	> 1000	N.D.	N.D.	> 100
Aspergillus floccosus	21405	21.01	84.65	N.D.	N.D.	100
Aspergillus japonicus	21425	12.29	69.06	N.D.	N.D.	> 100
Aspergillus urmiensis	21396	22.75	105.67	N.D.	N.D.	N.D.
Cladosporium sphaerospermum	21388	> 50	> 500	66.57	N.D.	N.D.
Lulwoana sp.	21398	> 50	> 1000	> 417	N.D.	N.D.
Paracamarosporium sp.	21423	> 50	> 500	N.D.	N.D.	> 100
Penicillium chrysogenum	21395	56.75	214.14	N.D.	N.D.	N.D.
Penicillium citrinum	21390	> 50	> 1000	96.06	N.D.	N.D.
Penicillium sp. 1	21380	> 50	> 500	N.D.	50.00	> 100
Penicillium sp. 1	21386	10.24	102.53	N.D.	12.50	> 100
Penicillium sp. 1	21387	16.34	91.48	N.D.	50.00	100
Penicillium sp. 1	21389	11.88	101.30	N.D.	< 6.25	25
Penicillium sumatrense	21382	> 50	> 1000	N.D.	N.D.	> 100
Phoma sp.	21426	6.31	40.41	N.D.	N.D.	N.D.
Pseudeurotium bakeri	21422	> 50	> 1000	N.D.	< 6.25	100
Talaromyces liani	21412	27.58	208.09	N.D.	> 100	> 100
Talaromyces sp. 1	21276	> 50	> 1000	N.D.	N.D.	> 100
Talaromyces sp. 3	21415	> 50	> 1000	N.D.	> 100	> 100
Talaromyces sp. 3	21421	> 50	> 1000	N.D.	> 100	50
Trichoderma harzianum	21394	> 50	> 500	N.D.	> 100	100
T. harzianum	21411	> 50	> 500	121.00	N.D.	> 100
Westerdykella capitulum	21407	> 50	N.D.	> 417	N.D.	> 100
Ascorbic acid *		13.70	6.80			N.D.
Kojic acid *				49.32		

KUC ID, Korea University Culture collection ID; ABTS, 2,2'-azino-bis-3-ethylbenzothiazoline-6-sulfonic acid; DPPH, 2,2-diphenyl-1-picrylhydrazyl; IC$_{50}$, half maximal inhibitory concentration; MIC, minimum inhibitory concentration; N.D., not detected. * positive controls.

A. japonicas KUC21396, *Penicillium* sp. 1 KUC21386 and KUC21389, and *Phoma* sp. KUC21426 extracts exhibited even higher ABTS radical-scavenging activity than the positive control, ascorbic acid. In particular, the IC_{50} values of both ABTS and DPPH radical-scavenging activity of *Phoma* sp. KUC21426 extract were significantly lower than the values of other extracts, implying its high antioxidative capacity. To the best of our knowledge, this is the second report of radical-scavenging activity of the genus *Phoma*, which contains a number of plant pathogen species.

2.2.2. Tyrosinase Inhibitory Activity

A total of five fungal extracts (*Cladosporium sphaerospermum* KUC21388, *Lulwoana* sp. KUC21398, *P. citrinum* KUC21390, *Trichoderma afroharzianum* KUC21411, and *Westerdykella capitulum* KUC21407) showed tyrosinase inhibitory activity (Table 5). Tyrosinase is able to oxidize L-3,4-dihydroxyphenylalanine (L-DOPA) to dopaquinone, which is eventually converted to pheomelanin [48]. As melanin biosynthesis is responsible for darkening of the skin tone, researchers have tried to discover tyrosinase inhibitors. Tyrosinase inhibitors can suppress melanin formation in the skin, so they can be developed as whitening agents. In particular, kojic acid, the most widely used tyrosinase inhibitor, was discovered through the screening of 600 marine fungi [49].

There is no report of tyrosinase inhibitors from the genera *Cladosporium*, *Lulwoana*, and *Westerdykella*. It has been discovered that various *Penicillium* spp. can produce kojic acid, but there has been no report of any tyrosinase inhibitor from *P. citrinum*. On the other hand, several tyrosinase inhibitors have been reported from the genus *Trichoderma*, for example, homothallin II from *T. viride* [50] and 1-(1,2,5-trihydroxy-3-isocyanopent-3-enyl)-ethanol, 1-(3-chloro-1,2-dihydroxy-4-isocyano-4-cyclopenten-1-yl)ethanol, and 1-(1,2,3-trihydroxy-3-isocyano-4-cyclopenten-1-yl)ethanol from *T. harzianum* [51,52]. Some antioxidants have been reported to be able to inhibit tyrosinase by scavenging reactive quinone products [48]. Since none of the five fungal extracts exhibited remarkable radical-scavenging activity, it is obvious that their tyrosinase inhibitors use inhibiting mechanisms other than scavenging reactive quinone products.

Considering that all the activities were measured using crude extracts, the tyrosinase inhibitory activity of *C. sphaerospermum* KUC21388, *P. citrinum* KUC21390, and *Trichoderma harzianum* KUC21411 extracts were notable. In particular, the IC_{50} value of *C. sphaerospermum* KUC21388 extract was even comparable to that of the positive control, kojic acid. This is the first report of tyrosinase inhibition activity in the genus *Cladosporium*, which is generally known to produce melanin.

2.2.3. Antifungal Activity

Antifungal experiments on *Asteromyces cruciatus* and *Lindra thalassiae* were conducted using the extracts of fungi isolated from intertidal mudflats and abandoned salterns. The two target fungi were selected considering that our fungal isolates were derived from the marine environment. *A. cruciatus* is a potentially harmful fungus to brown algae because it degrades alginate, a major constituent of brown algae. Alginate plays an important role in brown algae by forming the structure of the algal biomass and physically protecting algae from pathogens. *L. thalassiae* is a pathogenic fungus of sea plants and brown algae that causes raisin disease [53–55]. The results showed that the extracts of 17 fungal strains showed inhibitory effects against mycelial growth in *A. cruciatus* or *L. thalassiae* (Table 5).

All the extracts of the four *Penicillium* sp. 1 strains had potential antifungal activities, as they could inhibit the growth of *A. cruciatus* at a minimum concentration of 50 µg/mL and could inhibit the growth of *L. thalassiae* as well. Since they were identified as the same species from the same isolation source, the potential antifungal secondary metabolites are highly likely to be identical. There are a number of antifungal compounds reported from the genus *Penicillium*, for example, brefeldin A from *P. brefeldianum*; griseofulvin from *P. griseofulvum*; atpenins A4, A5, and B from *P. atramentosum* [56]; xanthocillin X and penicisteroid A from *P. chrysogenum* [56,57]; calbistrins from *P. restrictum* [58]; canadensolide from *P. canadense* [59]; macrocyclic polylactones from *P. verruculosum* [60]; patulin from *P. carneum* , citrinin from *P. melinii* (previously known as *P. damascenum*), palitantin and arthrographol from *P. implicatum* [61]; and compactin from *P. brevicompactum* [62]. There is no report of antifungal compounds or activity from the genus *Pseudeurotium*. In the case of the genus *Talaromyces*, some antifungal compounds have been reported, for example, talaroconvolutins from *T. convolutus* [63], talaron from *T. flavus* [64], 3-*O*-methylfunicone from *T. pinophilus*, wortmannin from *T. wortmannii*, macrophorin A from *T. purpurogenus*, and botryodiplodin from *T. stipitatus* [56]. In particular, *Penicillium* sp. 1 KUC21389 and *Pseudeurotium bakeri* KUC21422 exhibited significantly lower MIC values than the other strains. To the best of our knowledge, this is the first report of the antifungal activity of the genus *Pseudeurotium*. It is strongly speculated that these antifungal metabolites can affect other pathogenic fungi, so they could be developed as biocontrol agents.

2.2.4. Quorum Sensing Inhibitory Activity

Since only a few fungal quorum-sensing inhibitors (QSIs) have been found to date, it is worth investigating fungi from extreme environments such as intertidal mudflats and abandoned salterns. The two-step screening process showed that the extracts of *A. urmiensis* KUC21392 moderately inhibited the production of violacein by *C. violaceum* CV026 in the presence of N-(3-oxo-hexanoyl)-ʟ-homoserine lactone (3-oxo-C6-HSL), an N-acyl homoserine lactone (AHL), which is one of the most widely studied quorum sensing (QS) molecules (Figure 6). Bacterial biofilms cause huge economic losses in many areas, such as the food industry, aquaculture, wastewater treatment, and the shipping industry [65,66]. QS is responsible for biofilm formation and the expression of virulence genes, and bacterial biofilms have been reported to be approximately 1,000 times more resistant to antibiotics than their planktonic counterparts [67]. Some fungi produce quorum-quenching compounds such as patulin and penicillic acid from *Penicillium* spp., which can suppress bacterial biofilm formation [68]. Other studies reported that the extracts of *Fusarium graminearum*, *Lasidiplodia* sp., *Phellinus igniarius*, and fruiting bodies of *Auricularia auricular*, *Tremella fuciformis*, and *Ganoderma lucidum* have QS inhibitory activity [69–74]. Fungal QSIs were reported in the genera *Penicillium* and *Trichoderma*: polyhydroxyanthraquinones, penicillic acid, and patulin from *P. restrictum*, *P. radicicola*, and *P. coprobium*, respectively [68,75]; and carot-4-en-9,10-diol from *T. virens* [76]. *Penicillium* spp. are a good source of QSIs, considering that 33 out of 100 *Penicillium* extracts were screened as potential QSI producers in a previous study. In addition, *Penicillium atramentosum* was reported to produce QSIs other than penicillic acid or patulin [77]. *A. urmiensis* KUC21392 was identified as a potential QSI producer and is valuable as a source of natural compounds that regulate bacterial biofilms and related diseases. This is the first report on the QS inhibitory activity of the genus *Aspergillus*.

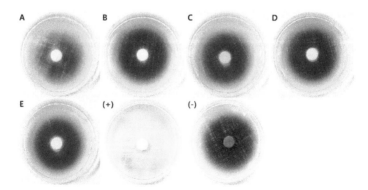

Figure 6. Quorum-sensing inhibitory activity of fungal extracts. (**A**) *Aspergillus urmiensis* KUC21392; (**B**) *Penicillium* sp. 2 KUC21400; (**C**) *Talaromyces stipitatus* KUC21402; (**D**) *Trichoderma afroharzianum* KUC21401; (**E**) *Trichoderma afroharzianum* KUC21404; (+) positive control (piericidin A); (−) negative control (DMSO).

Interestingly, four fungal extracts (*Penicillium* sp. 2 KUC21400, *T. stipitatus* KUC21402, *T. afroharzianum* KUC21401 and KUC21404) inhibited the growth of *C. violaceum*, resulting in a distinct hollow circle around the paper disc (Figure 5). Based on their antibacterial activity against *C. violaceum*, these four fungal strains were found to be producers of antimicrobial compounds that inhibit gram-negative bacteria. A number of antibacterial compounds have been discovered from the genus *Penicillium*, including rugulotrosin A and B, secalonic acid B and D, penicillixanthone A, citrinin, viridicatumtoxin B, penialidin B and C, penicyclones A–E [78–83]. *P. rubens* is the producer of penicillin, the first antibiotic that triggered the discovery of antibiotics [84]. Norlichexanthone, 3,1′-didehydro-3[2″(3′″,3′″-dimethyl-prop-2-enyl)-3″-indolylmethylene]-6-methyl pipera-zine-2,5-dione, gladiolic acid, and talaroderxines A and B were also produced by *Penicillium solitum*, *P. chrysogenum*, *Penicillium gladioli*, and *Penicillium derxii*, respectively [85–88]. The genus *Talaromyces*, which was formerly regarded as the teleomorphic state of *Penicillium*, has also been reported to produce various antibacterial compounds: fumitremorgin C, pseurotin A_1 and A_2, 3-dehydroxymethylbisdethio-3,10a-bis(methylthio)gliotoxin, bisdethiobis(methylthio)gliotoxin, didehydrobisdethiobis(methylthio)gliotoxin, 7-hydroxy-deoxytalaroflavone, deoxytalaroflavone, 7-epiaustdiol, 8-O-methylepiaustdiol, stemphyperylenol, skyrin, secalonic acid A, emodin, norlichexanthone, and talaromycesone A [89–93]; (−)luteoskyrin, cyclochlorotine, (+)rugulosin, rubroskyrin, lumiluteoskyrin, simotoxin, pibasterol, and skyrin produced from *T. islandicus* and *Talaromyces rugulosus* [94]; wortmannin, flavomannins A-D, and talaromannins A and B from *T. wortmannii* [95,96]; (−)-8-hydroxy-3-(4-hydroxypentyl)-3,4-dihydroisocoumarin, pentalsamonin, 5-hydroxy-7-methoxy-2-methylbenzofuran-3-carboxylic acid and 1-(5-hydroxy-7-methoxybenzofuran -3-yl)ethan-1-one, bacillosporin A, and talaromycolides A–C from *Talaromyces verruculosus*, *T. purpureogenus*, *Talaromyces amestolkiae*, *Talaromyces bacillisporus*, and *T. pinophilus* respectively [45,97–100]; talaraculone B, (-)mitorubrin, and bacillisporin A and B from *Talaromyces aculeatus* [101,102]; and 2,2′,3,5′-tetrahydroxy-3′-methylbenzophenone, 1,4,7-trihydroxy-6-methylxa nthone, and 1,4,5-trihydroxy-2-methylxanthone from *T. islandicus* [103]. In the case of *T. stipitatus*, talaromyone was reported as an antibacterial compound [104]. Therefore, the potential antibacterial compound produced by *T. stipitatus* KUC21402 is likely to be talaromyone. The genus *Trichoderma* is well known for producing various antibacterial compounds: 3-(3-oxocyclopent-1-enyl) propanoic acid, trichodermaol, alternariol 1′-hydroxy-9-methyl ether [105–107]; pseudokonin KL III and IV produced from *T. pseudokoningii* [108]; suzukacillin, viridian, lignoren, and gliotoxin from *T. viride* [109–112]; 6-n-pentyl-α-pyrone from *T. harzianum* and *T. longibrachiatum* [113]; shikimic acid from *T. ovalisporum* [114]; and coniothranthraquinone 1 and emodin from *T. aureoviride* [115]. Therefore, the potential antibacterial compounds of *T. afroharzianum* KUC21401 and KUC21404 are unlikely to be novel, but this is the first report on the antibacterial activity of this species.

3. Materials and Methods

3.1. Microorganisms and Diversity Analyses

3.1.1. Sediment Sampling and Fungal Isolation

Intertidal sediments 10 cm deep from the surface were sampled into sterile polyethylene jars using a sterile knife. The sediment samples were sieved using 2-mm sieves and stored at −80 °C for further use. To isolate fungi from the sediment samples, 1 gram of wet sediment was mixed with 10 mL of sterilized D.W. and vortexed. Preliminary tests with a variety of different intertidal sediment samples showed that the optimal concentration for isolating fungi from intertidal sediments was 5–10 times higher than those typically used when using forest sediments. The reason may be that the amount of organic matter in the sediment is relatively small, so that the amount of microbial biomass is small. The sediment suspension was serially diluted to a ratio of 1 g : 1000 mL, and 3 mL of the diluted suspension was added to 47 mL of the culture medium composed of 2% malt extract, 3.59% artificial sea salt (Instant Ocean, Aquarium Systems, Mentor, OH, USA), and 0.01% streptomycin sulfate (Sigma-Aldrich, St. Louis, MO, USA) for inhibiting the growth of bacteria. After that, 150 μL of the inoculated medium was distributed into each well of four sterilized 96-well plates. The plates were incubated at room temperature for 2–4 weeks until the fungal mycelia were visible to the naked eye. Each fungal colony was transferred to an individual 2% malt extract agar plate with 0.01% streptomycin sulfate. The plates were incubated at room temperature for a week and subcultured to isolate single strains. The isolated fungal strains were stored in 10% glycerol solution at 4 °C for further use and deposited in the Korea University Culture (KUC) collection.

3.1.2. DNA Extraction, PCR, and Identification

Genomic DNA was extracted from fungal culture using the *AccuPrep*® Genomic DNA Extraction Kit (Bioneer, Seoul, Republic of Korea), of which lysis buffer was replaced with fungal lysis buffer (12 mL of 0.5 M EDTA pH 8.0, 40 mL of 1 M Tris-HCL pH 8.0, 1 g of SDS, 0.8766 g of NaCl, and D.W. to a final volume of 100 mL). PCR amplification of ITS regions were carried out with ITS1F-LR3 primer [ITS1F (5′-CTT GGT CAT TTA GAG GAA GTA A-3′) and LR3 (5′- CCG TGT TTC AAG ACG GG-3′) [116,117] using the *AccuPower*® PCR Premix (Bioneer, Seoul, Republic of Korea). *benA* for *Talaromyces* species was amplified using the Bt2a-Bt2b primer [Bt2a (5′-GGT AAC CAA ATC GGT GCT GCT TTC-3′) and Bt2b (5′-ACC CTC AGT GTA GTG ACC CTT GGC-3′)] [118]. EF1-α for *Trichoderma* species was amplified using EF1-EF2 primer [EF1 (5′-ATG GGT AAG GAR GAC AAG AC-3′) and EF2 (5′-GGA RGT ACC AGT SAT CAT GTT-3′)] [119]. ITS regions were amplified under the conditions: 95 °C for 4 min, followed by 35 cycles of 95 °C for 30 s, 54 °C or 55 °C for 30 s, and 72 °C for 30 s. An elongation step of 72 °C for 5 min was performed at the end. For *benA*, the conditions were as follows: 95 °C for 5 min, followed by 38 cycles of 95 °C for 35 s, 55 °C or 56 °C for 50 s, and 72 °C for 2 min; an elongation step was performed at 72 °C for 7 min. For EF1-α, the conditions were as follows: 94 °C for 2 min, followed by 34 cycles of 93 °C for 30 s, 56 °C for 30 s, and 72 °C for 1 minute; and an elongation step was performed at 72 °C for 10 min. Through electrophoresis with 1% agarose gel, PCR products were conformed. PCR products were purified with the *AccuPrep*® PCR Purification Kit (Bioneer, Seoul, Republic of Korea) according to the user guide. DNA sequencing was performed using an 3730xl DNA Analyzer (Life technology, Carlsbad, CA, USA) with indicated PCR primers by Macrogen (Seoul, Republic of Korea). All the sequences were submitted to the GenBank and are presented in Table 1.

3.1.3. Phylogenetic Analysis

All the obtained sequences were assembled, proofread and edited using MEGA v. 7.0 [120]. Edited sequences were aligned using MAFFT 7.130 [121] and ambiguously aligned positions manually modified using MacClade 4.08 [122]. Through MrModeltest 2.3 using Akaike information criterion criteria, the best-fit models were calculated for the Bayesian phylogenetic analysis [123]. The Bayesian phylogenetic analyses were performed for 10 million generations using MrBayes 3.2.1 [124]. Phylogenetic trees were sampled every 100th generation. After sampling, the last 75% of the trees were selected. The constructed phylogenetic tree followed the 50% majority-rule and the tree reliability was presented by posterior probability.

3.1.4. DNA library Preparation and Amplicon Sequencing

Total DNA was extracted from 0.3 g of frozen sediment using a PowerSoil DNA isolation kit (MoBio, Carlsbad, CA, USA) following the manufacturer's protocol. DNA libraries were constructed using the Illumina MiSeq platform with the fungal ITS rDNA gene. An approximate 300–350 bp region of the ITS2 region was amplified with forward primer fITS7 [125] and reverse primer ITS4 [116]. Primer fITS7 contained a unique 12-nt barcode at the 5′ end for MiSeq sequencing detection. Paired-end read sequences were generated by high-throughput sequencing technology.

3.1.5. Bioinformatics Analyses

The sequences obtained in this study were processed using QIIME v1.9.1 [126]. The primer, key, and barcode sequences were trimmed from both ends. Sequences composed of homopolymers (n > 6) and chimeras were removed, and the remaining sequences were analyzed. The sequences were then clustered as operational taxonomic units (OTUs) based on a ≥ 97% similarity threshold and the average linkage method using Vsearch [127]. The representative sequence that was most abundant from each OTU was taxonomically assigned using the UNITE [128] reference database.

Nonmetric multidimensional scaling (NMDS) analysis applying the Bray–Curtis similarity index was performed to plot the similarity of the fungal communities in a way such that distances could be represented in two dimensions, using the vegan package in R statistical software, version 3.5.3. Alpha diversity indices were calculated using QIIME, and differences among the sediments from the abandoned salterns and mudflats were compared. After taxonomic profiling, we also investigated which fungal taxa differed in relative abundance among the intertidal environments. The relative abundances of the fungal communities were compared using linear discriminant analysis effect size (LefSe) analysis [129] with a LDA score threshold of 2.0 at the phylum, class, order, and family levels. Indicator species analysis was performed using the labdsv package in R to determine the key fungal taxa that best represent the Yellow Sea intertidal environments. Indicator species values are based on how specific and widespread the taxa are within a particular group and are independent of the relative abundance of other fungi [130]. RandomForest analysis was performed to identify key predictors of the environments among the fungal taxa. The accuracy importance measure was calculated for each tree and averaged over the forest (10,000 trees). This analysis was performed using the randomForest package in R. To determine which fungal taxa differed among the environments, the multivariate statistical package mvabund [131] was used in R; this approach is reported to have much greater power than distance-based approaches. In the generalized linear model, the relative abundance of each fungal taxon was modeled on the negative binomial distribution.

3.2. Preparation of Fungal Extracts

All the fungal isolates were precultured on 90-mm Petri dishes (10090, SPL Life Sciences Co., Pocheon, Republic of Korea) containing 20 mL of potato dextrose agar (PDA) at 25 °C for a week. After that, three agar plugs with actively-growing mycelia were inoculated to 150-mm Petri dishes (10150, SPL Life Sciences Co., Pocheon, Republic of Korea) containing 50 mL of PDA and incubated at 25 °C for 7 days in the dark. The cultures were extracted with 200 mL of MeOH and filtered with Whatman No.1 filter paper. The filtrates were dried at 35 °C using a rotary evaporator, and the dried extracts were dissolved in 15 mL of D.W. and 15 mL of EtOAc. After 6 hours, the EtOAc layers were transferred to 20-mL scintillation vials and dried using the rotary evaporator, and the dried extracts were stored at 4 °C until use. All the extracts were prepared in triplicate.

3.3. Biological Assays

The biological assays were performed with methods that were the same as or slightly modified from the methods used in our previous study [132].

3.3.1. Antioxidant Assay

ABTS Radical-Scavenging Assay

The 7-mM ABTS (Sigma-Aldrich, St. Louis, MO, USA) solution in phosphate-buffered saline (PBS, pH 7.4) was mixed with potassium persulfate solution that dissolved in PBS to 2.45 mM. The mixture was stored in the dark at room temperature for 24 h to form the ABTS$^{\bullet+}$. The solution was diluted with PBS to an absorbance of 0.70 (±0.02) at a wavelength of 734 nm. Then, 198 μL of ABTS radical solution was mixed with 2 μL of the fungal extract (10 mg/mL DMSO) in a 96-well plate. The absorbance was measured at 734 nm after 6 min using a spectrophotometer (Sunrise^{TM}, Tecan

Group Ltd., Port Melbourne, VIC, Australia). Ascorbic acid (Sigma Aldrich, St. Louis, MO, USA) was used as a positive control.

DPPH Radical-Scavenging Assay

The 200 μL of 150 μM DPPH (Sigma-Aldrich, St. Louis, MO, USA) solution in 80% methanol was mixed with 22 μL of the fungal extract (10 mg/mL DMSO) in a 96-well plate. The plate was stored in the dark at room temperature, and the absorbance was measured at 540 nm after 30 min using the spectrophotometer. Ascorbic acid was used as a positive control.

3.3.2. Tyrosinase Inhibition Assay

Tyrosinase inhibitory activity was measured based on the method of Lai et al. (2009) [133]. Briefly, 70 μL of 0.1 M potassium phosphate buffer (pH 6.8), 40 μL of fungal extract (2.5 mg/mL 50% DMSO), and 30 μL of tyrosinase from mushroom (0.02 mg/mL buffer; Sigma Aldrich, St. Louis, MO, USA) were added into each well of the 96-well plate. The plates were floated on water in a waterbath at 30 °C for 5 minutes, and 100 μL of 2.5 mM L-DOPA (Sigma Aldrich, St. Louis, MO, USA) was added. After 30 min, the plates were placed in ice to terminate the reaction, and the absorbance was measured at 492 nm. Kojic acid (Sigma Aldrich, St. Louis, MO, USA) was used as a positive control.

3.3.3. Antifungal Assay

Asteromyces cruciatus SFC20161110-M19 was obtained from Marine Fungal Resource Bank (MFRB) at Seoul National University as a marine bioresource bank of Korea by the Ministry of Oceans and Fisheries, and *Lindra thalassiae* NBRC106646 was purchased from Biological Resource Center under National Institute of Technology and Evaluation. Antifungal activity was determined in a 96-well plate using the microtiter broth dilution method [134]. Twenty-five microliters of spore suspensions (4×10^5 conidia/mL) of the target fungi were added to each well containing 25 μL of 4X potato dextrose broth and 49 μL of D.W. The fungal extracts were added to a final concentration of 100 μg/mL. The inoculated plates were incubated at 25 °C for 3 days. The extracts with lower concentrations (50, 25, 12.5, 6.5 μg/mL) were tested to determine MIC, the minimum concentration of an antimicrobial agent that causes microbial death.

3.3.4. Quorum Sensing Inhibition Assay

QSI screening was performed based on the inhibition of violacein production by *Chromobacterium violaceum* CV026 strains under culture conditions supplemented with an exogenous QS molecule, 3-oxo-C6-HSL [135]. The *C. violaceum* CV026 cultured overnight was diluted with LB medium (5 g of yeast extract, 10 g of tryptone, and 5 g of NaCl in a liter of D.W.) to an $OD_{600 \, nm}$ of 0.1, and 97 μL of the diluted culture was added to each well of a 96-well plate. One microliter of 3-oxo-C6-HSL (final concentration of 10 μM; Sigma–Aldrich, St. Louis, MO, USA) and 2 microliters of fungal extract (10 mg/mL DMSO) were added to each well. As a negative and positive control, two microliters of DMSO and two microliters of 100 nM (Z)-4-bromo-5-(bromomethylene)-2(5H)-furanone (Furanone C-30; Sigma-Aldrich, St. Louis, MO, USA) were used, respectively. After 16 h at 28 °C, 100 μL of DMSO was added to each well and vigorously shaken for an hour at room temperature to determine the production of violacein.

The selected extracts were further examined using a paper disc method. Fifteen microliters of the diluted culture of *C. violaceum* CV026 was spread on an LB agar plate, and a paper disc (diameter 8 mm; Advantec, Tokyo, Japan) impregnated with 40 μL of extract and 1 μL of 1 mM 3-oxo-C6-HSL was placed in the center of the plate. DMSO was used as a negative control, and piericidin A isolated from *Streptomyces xanthocidicus* KPP01532 was used as a positive control [135]. The plates were incubated overnight at 28 °C.

3.4. Statistical Analyses

The half maximal inhibitory concentration (IC_{50}) values were calculated by nonlinear regression analysis using SigmaPlot 12.0 (Systat Software Inc., San Jose, CA, USA).

4. Conclusions

The results of the fungal community analysis showed that only the abandoned saltern in Yongyudo was significantly different from the other abandoned salterns or mudflats. The fungal communities in the abandoned salterns in Yubudo and Gopado had been restored to a similar status as those in the mudflats, but that was not the case of the saltern in Yongyudo, which was abandoned less than a year ago. The results also implied that it takes more than 35 years to restore abandoned salterns to mudflats via natural restoration. The α-diversity of the fungal community in the NS was significantly lower than that of the RSs and mudflats. The diversity indices were high in general, which implies that these unique environments are valuable for bioprospecting.

Through various statistical analyses, the lineages of Gigasporales sp. and *Umbelopsis* sp. were selected as indicator taxa of mudflats in the Yellow Sea in South Korea. We demonstrate that the ecological recovery of the abandoned salterns can be assessed by using the two indicator taxa to indicate not only the recovery state but also the recovery direction, i.e., whether the recovery process is moving in the right direction. In this study, several analytical methods were applied to investigate indicator taxa. However, the results were not identical, even though we only used methods that use the absolute abundance data rather than processed data such as ecological distances that could lead to statistical misinterpretation [27]. Therefore, indicator taxa must be selected by carefully comparing the results of various types of analysis. We also proved that the fungal community is highly suitable for use in ecosystem assessment, as it has a sufficient level of identification with good resolution. The limitation was that the number of samples, especially for the NS and RS environments, was insufficient. A larger sample size is recommended for more accurate statistical analysis with high fidelity.

To evaluate the fungal diversity and compare it with the results from the metagenome analysis, the culturable fungi were isolated from the sediment samples. A total of 53 fungal strains was isolated, and the number of fungal isolates differed by sampling location rather than by environment type. Based on the phylogenetic analysis, *Talaromyces* was the most diverse genus, consisting of seven different species. The genera *Talaromyces* and *Trichoderma* were mostly isolated from the mudflat sediments, and most *Penicillium* spp. were found in the NS. The diversity and the number of isolates in the NS were higher than those in the other salterns, which is inconsistent with the results of metagenomics, and every isolate except *P. chrysosporium* belongs to the Ascomycota. This result showed that metagenomics must be applied when examining the microbial diversity within an actual environment, as has been suggested over the past decades [29]. Interestingly, a total of 18 fungal isolates was identified as candidate novel species.

To evaluate the potential value of the intertidal environments for bioprospecting, a variety of biological activities of the fungal extracts were investigated. Regardless of the actual diversity, the isolates from both intertidal environments exhibited a variety of biological activities. Over half of the culturable fungal isolates (29 strains) exhibited biological activities, and *Penicillium* spp. exhibited the most varied activities. The antioxidant compound of *Phoma* sp. KUC21426, the tyrosinase inhibitors of *C. sphaerospermum* KUC21388, the antifungal compounds of *Penicillium* sp. 1 KUC21389 and *P. bakeri* KUC21422, and the quorum-sensing inhibitor of *A. urmiensis* KUC21392 will be separated and identified in the near future. Since bioactive compounds with these activities have never been reported from these species, the likelihood of discovering novel compounds is high.

In this study, we presented the fungal community and diversity of the abandoned solar salterns and the intertidal mudflats on three different islands located in the Yellow Sea of South Korea, and we identified indicator taxa for the assessment of the restoration of abandoned salterns to mudflats. Additionally, we provided reliable DNA information for the 53 fungal isolates from the intertidal sediments and their exploitable biological activities. We demonstrated that these unique environments are highly valuable in bioprospecting not only for novel microorganisms but also for novel bioactive compounds. Restoring the reduced biodiversity of abandoned salterns is important in terms of environmental protection and conservation. However, many novel fungal species were found from the abandoned salterns, and interestingly, all fungal strains selected for high bioactivity were isolated from NS or RS. Therefore, active measures to recover abandoned salterns to mudflats should be carefully determined and planned, taking into account their value as a source of biological resources.

Supplementary Materials: The following are available online at http://www.mdpi.com/1660-3397/17/11/601/s1, Figure S1: Linear discriminant analysis (LDA) scores of the fungal taxa of which score > 2. NS and RS mean non-recovered saltern and recovered saltern, respectively, Figure S2: Relative abundance of Ascomycota in the three different intertidal environments. NS and RS mean non-recovered saltern and recovered saltern, respectively, Figure S3: The Bayesian analysis tree based on EF1-α for *Trichoderma* complex. Bayesian posterior probabilities (PP) at the nodes are presented if > 75. All of reference strains are type and ex-type strains. The fungi isolated from this study are in bold. The scale bar means the number of nucleotide substitutions per position, Figure S4: The Bayesian analysis tree based on combined of ITS and *benA* for *Talaromyces* complex. Bayesian posterior probabilities (PP) at the nodes are presented if > 75. Type strains are indicated as *. The fungi isolated from this study are in bold. The scale bar means the number of nucleotide substitutions per position, Table S1: The α-diversities of culturable fungal isolates from abandoned salterns and intertidal mudflats.

Author Contributions: Conceptualization, Y.M.H., H.L., and J.-J.K.; Methodology, Y.M.H., H.L., S.L.K., M.Y.P., and J.E.K.; Validation, G.-H.K., B.S.K., and J.-J.K.; Formal Analysis, Y.M.H., K.K., and M.Y.P.; Investigation, Y.M.H., K.K., and S.L.K.; Resources, G.-H.K. and J.-J.K.; Writing-Original Preparation, Y.M.H., H.L., K.K., S.L.K., M.Y.P., and J.E.K.; Writing-Review & Editing, H.L., G.-H.K., B.S.K., and J.-J.K.; Visualization, Y.M.H., H.L., and S.L.K.; Supervision, G.-H.K., B.S.K., and J.-J.K.; Project Administration, J.-J.K.; Funding Acquisition, J.-J.K..

Funding: This research was funded by National Research Foundation of Korea (NRF) funded by the Korean government (MSIT) grant number 2017R1A2B4002071 and "Development of blue carbon information system and its assessment for management (20170318)" funded by the Ministry of Oceans and Fisheries of Korea (MOF) granted to JSK, JR, and BOK.

Conflicts of Interest: The authors declare no conflict of interest.

References

1. Zhao, D.L.; Wang, D.; Tian, X.Y.; Cao, F.; Li, Y.Q.; Zhang, C.S. Anti-phytopathogenic and cytotoxic activities of crude extracts and secondary metabolites of marine-derived fungi. *Mar. Drugs* **2018**, *16*, 36. [CrossRef] [PubMed]

2. Sridhar, K. Marine filamentous fungi: Diversity, distribution and bioprospecting. In *Developments in Fungal Biology and Applied Mycology*; Springer: Singapore, 2017; pp. 59–73.

3. Saravanakumar, K.; Yu, C.; Dou, K.; Wang, M.; Li, Y.; Chen, J. Biodiversity of *Trichoderma* community in the tidal flats and wetland of southeastern China. *PLoS ONE* **2016**, *11*, e0168020. [CrossRef] [PubMed]

4. Park, C.K.; Oh, B.H.; Lee, D.W. Hydraulic characteristics of the non-power soil cleaning and keeping system by the large-scale model test at the dike gate. *J. Korean Soc. Agric. Eng.* **2014**, *56*, 67–75.

5. Park, N. *The Eco-Economical Evaluation of Tidal Flat Restoration in Suncheon Bay*; Pukyong National University: Busan, Korea, 2018.

6. Kim, H.S. On the value of coastal wetlands created from conservation and restoration of tidal flat ecosystem. *Korean Soc. Civ. Eng. Mag.* **2018**, *66*, 8–9.

7. Chávez, R.; Fierro, F.; García-Rico, R.O.; Vaca, I. Filamentous fungi from extreme environments as a promising source of novel bioactive secondary metabolites. *Front. Microbiol.* **2015**, *6*, 903. [CrossRef]

8. Hong, J.H.; Jang, S.; Heo, Y.M.; Min, M.; Lee, H.; Lee, Y.M.; Lee, H.; Kim, J.J. Investigation of marine-derived fungal diversity and their exploitable biological activities. *Mar. Drugs* **2015**, *13*, 4137–4155. [CrossRef]

9. Antón, J.; Rosselló-Mora, R.; Rodríguez-Valera, F.; Amann, R. Extremely halophilic bacteria in crystallizer ponds from solar salterns. *Appl. Environ. Microbiol.* **2000**, *66*, 3052–3057. [CrossRef]

10. Ballav, S.; Kerkar, S.; Thomas, S.; Augustine, N. Halophilic and halotolerant actinomycetes from a marine saltern of Goa, India producing anti-bacterial metabolites. *J. Biosci. Bioeng.* **2015**, *119*, 323–330. [CrossRef]

11. Chen, Q.; Zhao, Q.; Li, J.; Jian, S.; Ren, H. Mangrove succession enriches the sediment microbial community in South China. *Sci. Rep.* **2016**, *6*, 27468. [CrossRef]

12. Thatoi, H.; Behera, B.C.; Mishra, R.R.; Dutta, S.K. Biodiversity and biotechnological potential of microorganisms from mangrove ecosystems: A review. *Ann. Microbiol.* **2013**, *63*, 1–19. [CrossRef]

13. Lee, S.; Park, M.S.; Lim, Y.W. Diversity of marine-derived *Aspergillus* from tidal mudflats and sea sand in Korea. *Mycobiology* **2016**, *44*, 237–247. [CrossRef] [PubMed]

14. Sonjak, S.; Udovič, M.; Wraber, T.; Likar, M.; Regvar, M. Diversity of halophytes and identification of arbuscular mycorrhizal fungi colonising their roots in an abandoned and sustained part of Sečovlje salterns. *Soil Biol. Biochem.* **2009**, *41*, 1847–1856. [CrossRef]

15. Lee, H.; Lee, D.W.; Kwon, S.L.; Heo, Y.M.; Jang, S.; Kwon, B.O.; Khim, J.S.; Kim, G.H.; Kim, J.J. Importance of functional diversity in assessing the recovery of the microbial community after the Hebei Spirit oil spill in Korea. *Environ. Int.* **2019**, *128*, 89–94. [CrossRef] [PubMed]

16. Yun, K.; Khong, T.T.; Leutou, A.S.; Kim, G.D.; Hong, J.; Lee, C.H.; Son, B.W. Cristazine, a new cytotoxic dioxopiperazine alkaloid from the mudflat-sediment-derived fungus *Chaetomium cristatum*. *Chem. Pharm. Bull.* **2016**, *64*, 59–62. [CrossRef]

17. Nenkep, V.; Yun, K.; Son, B.W. Oxysporizoline, an antibacterial polycyclic quinazoline alkaloid from the marine-mudflat-derived fungus *Fusarium oxysporum*. *J. Antibiot.* **2016**, *69*, 709–711. [CrossRef]

18. Leutou, A.S.; Yun, K.; Son, B.W. Induced production of 6, 9-dibromoflavasperone, a new radical scavenging naphthopyranone in the marine-mudflat-derived fungus *Aspergillus niger*. *Arch. Pharmacal Res.* **2016**, *39*, 806–810. [CrossRef]

19. Yun, K.; Kondempudi, C.M.; Leutou, A.S.; Son, B.W. New production of a monoterpene glycoside, 1-O-(α-D-mannopyranosyl) geraniol, by the marine-derived fungus *Thielavia hyalocarpa*. *Bull. Korean Chem. Soc.* **2015**, *36*, 2391–2393. [CrossRef]

20. Leutou, A.S.; Yun, K.; Son, B.W. New production of antibacterial polycyclic quinazoline alkaloid, thielaviazoline, from anthranilic acid by the marine-mudflat-derived fungus *Thielavia sp.* *Nat. Prod. Sci.* **2016**, *22*, 216–219. [CrossRef]

21. Yun, K.; Leutou, A.S.; Rho, J.R.; Son, B.W. Formoxazine, a new pyrrolooxazine, and two amines from the marine–mudflat-derived fungus *Paecilomyces formosus*. *Bull. Korean Chem. Soc.* **2016**, *37*, 103–104. [CrossRef]

22. Bernhard, A.E.; Marshall, D.; Yiannos, L. Increased variability of microbial communities in restored salt marshes nearly 30 years after tidal flow restoration. *Estuar. Coasts* **2012**, *35*, 1049–1059. [CrossRef]

23. Meng, Q.; Yang, W.; Men, M.; Bello, A.; Xu, X.; Xu, B.; Deng, L.; Jiang, X.; Sheng, S.; Wu, X. Microbial community succession and response to environmental variables during cow manure and corn straw composting. *Front. Microbiol.* **2019**, *10*, 529. [CrossRef] [PubMed]

24. Cleary, M.; Oskay, F.; Doğmuş, H.T.; Lehtijärvi, A.; Woodward, S.; Vettraino, A.M. Cryptic Risks to Forest Biosecurity Associated with the Global Movement of Commercial Seed. *Forests* **2019**, *10*, 459. [CrossRef]

25. Dang, P.; Yu, X.; Le, H.; Liu, J.; Shen, Z.; Zhao, Z. Effects of stand age and soil properties on soil bacterial and fungal community composition in Chinese pine plantations on the Loess Plateau. *PLoS ONE* **2017**, *12*, e0186501. [CrossRef] [PubMed]

26. Wu, J.; Yu, M.; Xu, J.; Du, J.; Ji, F.; Dong, F.; Li, X.; Shi, J. Impact of transgenic wheat with wheat yellow mosaic virus resistance on microbial community diversity and enzyme activity in rhizosphere soil. *PLoS ONE* **2014**, *9*, e98394. [CrossRef] [PubMed]

27. Warton, D.I.; Wright, S.T.; Wang, Y. Distance-based multivariate analyses confound location and dispersion effects. *Methods Ecol. Evol.* **2012**, *3*, 89–101. [CrossRef]

28. Bücking, H.; Liepold, E.; Ambilwade, P. The role of the mycorrhizal symbiosis in nutrient uptake of plants and the regulatory mechanisms underlying these transport processes. In *Plant Science*; Dhal, N.K., Sahu, S.C., Eds.; Books on Demand: Norderstedt, Germany, 2012; Volume 4, pp. 107–138.

29. Streit, W.R.; Schmitz, R.A. Metagenomics—The key to the uncultured microbes. *Curr. Opin. Microbiol.* **2004**, *7*, 492–498. [CrossRef] [PubMed]

30. Huang, W.Y.; Cai, Y.Z.; Xing, J.; Corke, H.; Sun, M. A potential antioxidant resource: Endophytic fungi from medicinal plants. *Econ. Bot.* **2007**, *61*, 14–30. [CrossRef]

31. Amens, B. Dietary carcinogens and anticarcinogens. *Science* **1983**, *221*, 53.

32. Yen, G.C.; Chang, Y.C.; Sheu, F.; Chiang, H.C. Isolation and characterization of antioxidant compounds from *Aspergillus candidus* broth filtrate. *J. Agric. Food Chem.* **2001**, *49*, 1426–1431. [CrossRef]

33. Elaasser, M.M.; Abdel-Aziz, M.M.; El-Kassas, R.A. Antioxidant, antimicrobial, antiviral and antitumor activities of pyranone derivative obtained from *Aspergillus candidus*. *J. Microbiol. Biotechnol. Res.* **2017**, *1*, 5–17.

34. Kumar, C.G.; Mongolla, P.; Pombala, S.; Kamle, A.; Joseph, J. Physicochemical characterization and antioxidant activity of melanin from a novel strain of *Aspergillus bridgeri* ICTF-201. *Lett. Appl. Microbiol.* **2011**, *53*, 350–358. [CrossRef] [PubMed]

35. Li, X.; Li, X.M.; Xu, G.M.; Li, C.S.; Wang, B.G. Antioxidant metabolites from marine alga-derived fungus *Aspergillus wentii* EN-48. *Phytochem. Lett.* **2014**, *7*, 120–123. [CrossRef]

36. Ishikawa, Y.; Morimoto, K.; Iseki, S. Atrovenetin as a potent antioxidant compound from *Penicillium* species. *J. Am. Oil Chem. Soc.* **1991**, *68*, 666–668. [CrossRef]

37. Narasimhachari, N.; Gopalkrishnan, K.; Haskins, R.; Vining, L. The production of the antibiotic atrovenetin by a strain of *Penicillium herquei* Bainier & Sartory. *Can. J. Microbiol.* **1963**, *9*, 134–136.

38. Neill, K.; Raistrick, H. Studies in the biochemistry of micro-organisms, doi:100. Metabolites of *Penicillium atrovenetum* G. Smith. Part I. Atrovenetin, a new crystalline colouring matter. *Biochem. J.* **1957**, *65*, 166. [CrossRef]

39. Chen, C.H.; Shaw, C.Y.; Chen, C.C.; Tsai, Y.C. 2, 3, 4-trimethyl-5, 7-dihydroxy-2, 3-dihydrobenzofuran, a novel antioxidant, from *Penicillium citrinum* F5. *J. Nat. Prod.* **2002**, *65*, 740–741. [CrossRef]

40. Hayashi, K.I.; Suzuki, K.; Kawaguchi, M.; Nakajima, T.; Suzuki, T.; Numata, M.; Nakamura, T. Isolation of an antioxidant from *Penicillium roquefortii* IFO 5956. *Biosci. Biotechnol. Biochem.* **1995**, *59*, 319–320. [CrossRef]

41. Jakovljević, V.D.; Milićević, J.M.; Stojanović, J.D.; Solujić, S.R.; Vrvić, M.M. Antioxidant activity of ethanolic extract of *Penicillium chrysogenum* and *Penicillium fumiculosum*. *Hem. Ind.* **2014**, *68*, 43–49. [CrossRef]

42. Bringmann, G.; Lang, G.; Gulder, T.A.; Tsuruta, H.; Mühlbacher, J.; Maksimenka, K.; Steffens, S.; Schaumann, K.; Stöhr, R.; Wiese, J.; et al. The first sorbicillinoid alkaloids, the antileukemic sorbicillactones A and B, from a sponge-derived *Penicillium chrysogenum* strain. *Tetrahedron* **2005**, *61*, 7252–7265. [CrossRef]

43. Nenkep, V.N.; Yun, K.; Li, Y.; Choi, H.D.; Kang, J.S.; Son, B.W. New production of haloquinones, bromochlorogentisylquinones A and B, by a halide salt from a marine isolate of the fungus *Phoma herbarum*. *J. Antibiot.* **2010**, *63*, 199–201. [CrossRef] [PubMed]

44. Li, H.L.; Li, X.M.; Li, X.; Wang, C.Y.; Liu, H.; Kassack, M.U.; Meng, L.H.; Wang, B.G. Antioxidant hydroanthraquinones from the marine algal-derived endophytic fungus *Talaromyces islandicus* EN-501. *J. Nat. Prod.* **2017**, *80*, 162–168. [CrossRef]

45. Pandit, S.G.; Puttananjaih, M.H.; Harohally, N.V.; Dhale, M.A. Functional attributes of a new molecule-2-hydroxymethyl-benzoic acid 2'-hydroxy-tetradecyl ester isolated from *Talaromyces purpureogenus* CFRM02. *Food Chem.* **2018**, *255*, 89–96. [CrossRef] [PubMed]

46. Guo, J.; Ran, H.; Zeng, J.; Liu, D.; Xin, Z. Tafuketide, a phylogeny-guided discovery of a new polyketide from *Talaromyces funiculosus* Salicorn 58. *Appl. Microbiol. Biotechnol.* **2016**, *100*, 5323–5338. [CrossRef] [PubMed]

47. Yuan, W.H.; Teng, M.T.; Sun, S.S.; Ma, L.; Yuan, B.; Ren, Q.; Zhang, P. Active Metabolites from Endolichenic Fungus *Talaromyces* sp. *Chem. Biodivers.* **2018**, *15*, e1800371. [CrossRef] [PubMed]

48. Chang, T.S. An updated review of tyrosinase inhibitors. *Int. J. Mol. Sci.* **2009**, *10*, 2440–2475. [CrossRef] [PubMed]

49. Balboa, E.M.; Conde, E.; Soto, M.L.; Pérez-Armada, L.; Domínguez, H. Cosmetics from marine sources. In *Springer Handbook of Marine Biotechnology*; Springer: Berlin/Heidelberg, Germany, 2015; pp. 1015–1042.

50. Tsuchiya, T.; Yamada, K.; Minoura, K.; Miyamoto, K.; Usami, Y.; Kobayashi, T.; Hamada-Sato, N.; Imada, C.; Tsujibo, H. Purification and determination of the chemical structure of the tyrosinase inhibitor produced by *Trichoderma viride* strain H1-7 from a marine environment. *Biol. Pharm. Bull.* **2008**, *31*, 1618–1620. [CrossRef]

51. Lee, C.H.; Koshino, H.; Chung, M.C.; Lee, H.J.; Kho, Y.H. MR304A, a new melanin synthesis inhibitor produced by *Trichoderma harzianum*. *J. Antibiot.* **1995**, *48*, 1168–1170. [CrossRef]

52. Lee, C.H.; Koshino, H.; Chung, M.C.; Lee, H.J.; HONG, J.K.; Yoon, J.S.; Kho, Y.H. MR566A and MR566B, new melanin synthesis inhibitors produced by *Trichoderma harzianum*. *J. Antibiot.* **1997**, *50*, 474–478. [CrossRef]

53. Raghukumar, S. *Fungi in Coastal and Oceanic Marine Ecosystems*; Springer: New York, NY, USA, 2017.

54. Skriptsova, A.V. Fucoidans of brown algae: Biosynthesis, localization, and physiological role in thallus. *Russ. J. Mar. Biol.* **2015**, *41*, 145–156. [CrossRef]

55. Lane, A.L.; Mular, L.; Drenkard, E.J.; Shearer, T.L.; Engel, S.; Fredericq, S.; Fairchild, C.R.; Prudhomme, J.; Le Roch, K.; Hay, M.E.; et al. Ecological leads for natural product discovery: Novel sesquiterpene hydroquinones from the red macroalga *Peyssonnelia* sp. *Tetrahedron* **2010**, *66*, 455–461. [CrossRef]

56. Bladt, T.; Frisvad, J.; Knudsen, P.; Larsen, T. Anticancer and antifungal compounds from *Aspergillus*, *Penicillium* and other filamentous fungi. *Molecules* **2013**, *18*, 11338–11376. [CrossRef] [PubMed]

57. Gao, S.S.; Li, X.M.; Li, C.S.; Proksch, P.; Wang, B.G. Penicisteroids A and B, antifungal and cytotoxic polyoxygenated steroids from the marine alga-derived endophytic fungus *Penicillium chrysogenum* QEN-24S. *Bioorg. Med. Chem. Lett.* **2011**, *21*, 2894–2897. [CrossRef] [PubMed]

58. Jackson, M.; Karwowski, J.P.; Humphrey, P.E.; Kohl, W.L.; Barlow, G.J.; Tanaka, S.K. Calbistrins, novel antifungal agents produced by *Penicillium restrictum*. *J. Antibiot.* **1993**, *46*, 34–38. [CrossRef] [PubMed]

59. McCorkindale, N.J.; Wright, J.L.C.; Brian, P.W.; Clarke, S.M.; Hutchinson, S.A. Canadensolide—An antifungal metabolite of *Penicillium canadense*. *Tetrahedron Lett.* **1968**, *9*, 727–730. [CrossRef]

60. Breinholt, J.; Jensen, G.W.; Nielsen, R.I.; Olsen, C.E.; Frisvad, J.C. Antifungal macrocyclic polylactones from *Penicillium verruculosum*. *J. Antibiot.* **1993**, *46*, 1101–1108. [CrossRef] [PubMed]

61. Yamaji, K.; Fukushi, Y.; Hashidoko, Y.; Yoshida, T.; Tahara, S. Characterization of antifungal metabolites produced by *Penicillium* species isolated from seeds of *Picea glehnii*. *J. Chem. Ecol.* **1999**, *25*, 1643–1653. [CrossRef]

62. Brown, A.G.; Smale, T.C.; King, T.J.; Hasenkamp, R.; Thompson, R.H. Crystal and molecular structure of compactin, a new antifungal metabolite from *Penicillium brevicompactum*. *J. Chem. Soc. Perkin Trans.* **1976**, *11*, 1165–1170. [CrossRef]

63. Suzuki, S.; Hosoe, T.; Nozawa, K.; Kawai, K.I.; Yaguchi, T.; Udagawa, S.I. Antifungal substances against pathogenic fungi, talaroconvolutins, from *Talaromyces convolutus*. *J. Nat. Prod.* **2000**, *63*, 768–772. [CrossRef]

64. Kim, K.K.A.; Fravel, D.R.; Papavizas, G. Glucose oxidase as the antifungal principle of talaron from *Talaromyces flavus*. *Can. J. Microbiol.* **1990**, *36*, 760–764. [CrossRef]

65. Skandamis, P.N.; Nychas, G.J.E. Quorum sensing in the context of food microbiology. *Appl. Environ. Microbiol.* **2012**, *78*, 5473–5482. [CrossRef]

66. Schultz, M.P.; Bendick, J.A.; Holm, E.R.; Hertel, W.M. Economic impact of biofouling on a naval surface ship. *Biofouling* **2011**, *27*, 87–98. [CrossRef] [PubMed]

67. Olson, M.E.; Ceri, H.; Morck, D.W.; Buret, A.G.; Read, R.R. Biofilm bacteria: Formation and comparative susceptibility to antibiotics. *Can. J. Vet. Res.* **2002**, *66*, 86–92. [PubMed]

68. Rasmussen, T.B.; Skindersoe, M.E.; Bjarnsholt, T.; Phipps, R.K.; Christensen, K.B.; Jensen, P.O.; Andersen, J.B.; Koch, B.; Larsen, T.O.; Hentzer, M. Identity and effects of quorum-sensing inhibitors produced by *Penicillium* species. *Microbiology* **2005**, *151*, 1325–1340. [CrossRef] [PubMed]

69. Rajesh, P.; Rai, V.R. Hydrolytic enzymes and quorum sensing inhibitors from endophytic fungi of *Ventilago madraspatana* Gaertn. *Biocatal. Agric. Biotechnol.* **2013**, *2*, 120–124. [CrossRef]

70. Zhu, H.; Liu, W.; Wang, S.X.; Tian, B.Z.; Zhang, S.S. Evaluation of anti-quorum-sensing activity of fermentation metabolites from different strains of a medicinal mushroom, *Phellinus igniarius*. *Chemotherapy* **2012**, *58*, 195–199. [CrossRef] [PubMed]

71. Zhu, H.; He, C.C.; Chu, Q.H. Inhibition of quorum sensing in *Chromobacterium violaceum* by pigments extracted from *Auricularia auricular*. *Lett. Appl. Microbiol.* **2011**, *52*, 269–274. [CrossRef]

72. Zhu, H.; Sun, S. Inhibition of bacterial quorum sensing-regulated behaviors by *Tremella fuciformis* extract. *Curr. Microbiol.* **2008**, *57*, 418–422. [CrossRef]

73. Zhu, H.; Liu, W.; Tian, B.; Liu, H.; Ning, S. Inhibition of quorum sensing in the opportunistic pathogenic bacterium *Chromobacterium violaceum* by an extract from fruiting bodies of lingzhi or reishi medicinal mushroom, *Ganoderma lucidum* (w. Curt.: Fr.) p. Karst. (higher basidiomycetes). *Int. J. Med. Mushrooms* **2011**, *13*, 559–564. [CrossRef]

74. Sharma, R.; Jangid, K. Fungal quorum sensing inhibitors. In *Quorum Sensing Vs Quorum Quenching: A Battle with no End in Sight*; Springer: New Delhi, India, 2015; pp. 237–257.

75. Figueroa, M.; Jarmusch, A.K.; Raja, H.A.; El-Elimat, T.; Kavanaugh, J.S.; Horswill, A.R.; Cooks, R.G.; Cech, N.B.; Oberlies, N.H. Polyhydroxyanthraquinones as quorum sensing inhibitors from the guttates of *Penicillium restrictum* and their analysis by desorption electrospray ionization mass spectrometry. *J. Nat. Prod.* **2014**, *77*, 1351–1358. [CrossRef]

76. Wang, M.; Hashimoto, M.; Hashidoko, Y. Repression of tropolone production and induction of a *Burkholderia plantarii* pseudo-biofilm by carot-4-en-9, 10-diol, a cell-to-cell signaling disrupter produced by *Trichoderma virens*. *PLoS ONE* **2013**, *8*, e78024. [CrossRef]

77. Wang, L.; Zou, S.; Yin, S.; Liu, H.; Yu, W.; Gong, Q. Construction of an effective screening system for detection of *Pseudomonas aeruginosa* quorum sensing inhibitors and its application in bioautographic thin-layer chromatography. *Biotechnol. Lett.* **2011**, *33*, 1381–1387. [CrossRef] [PubMed]

78. Stewart, M.; Capon, R.J.; White, J.M.; Lacey, E.; Tennant, S.; Gill, J.H.; Shaddock, M.P. Rugulotrosins A and B: Two new antibacterial metabolites from an Australian isolate of a *Penicillium* sp. *J. Nat. Prod.* **2004**, *67*, 728–730. [CrossRef] [PubMed]

79. Bao, J.; Sun, Y.L.; Zhang, X.Y.; Han, Z.; Gao, H.C.; He, F.; Qian, P.Y.; Qi, S.H. Antifouling and antibacterial polyketides from marine gorgonian coral-associated fungus *Penicillium* sp. SCSGAF 0023. *J. Antibiot.* **2013**, *66*, 219–223. [CrossRef] [PubMed]

80. Subramani, R.; Kumar, R.; Prasad, P.; Aalbersberg, W. Cytotoxic and antibacterial substances against multi-drug resistant pathogens from marine sponge symbiont: Citrinin, a secondary metabolite of *Penicillium* sp. *Asian Pac. J. Trop. Biomed.* **2013**, *3*, 291–296. [CrossRef]

81. Zheng, C.J.; Yu, H.E.; Kim, E.H.; Kim, W.G. Viridicatumtoxin B, a new anti-MRSA agent from *Penicillium* sp. FR11. *J. Antibiot.* **2008**, *61*, 633–637. [CrossRef]

82. Jouda, J.B.; Kusari, S.; Lamshöft, M.; Talontsi, F.M.; Meli, C.D.; Wandji, J.; Spiteller, M. Penialidins A–C with strong antibacterial activities from *Penicillium* sp. an endophytic fungus harboring leaves of *Garcinia nobilis*. *Fitoterapia* **2014**, *98*, 209–214. [CrossRef]

83. Guo, W.; Zhang, Z.; Zhu, T.; Gu, Q.; Li, D. Penicyclones A–E, antibacterial polyketides from the deep-sea-derived fungus *Penicillium* sp. F23-2. *J. Nat. Prod.* **2015**, *78*, 2699–2703. [CrossRef]

84. Fleming, A. On the antibacterial action of cultures of a penicillium, with special reference to their use in the isolation of *B. influenzae*. *Br. J. Exp. Pathol.* **1929**, *10*, 226–236. [CrossRef]

85. Broadbent, D.; Mabelis, R.P.; Spencer, H. 3,6,8-Trihydroxy-1-methylxanthone—An antibacterial metabolite from *Penicillium patulum*. *Phytochemistry* **1975**, *14*, 2082–2083. [CrossRef]

86. Devi, P.; Rodrigues, C.; Naik, C.; D'Souza, L. Isolation and characterization of antibacterial compound from a mangrove-endophytic fungus, *Penicillium chrysogenum* MTCC 5108. *Indian J. Microbiol.* **2012**, *52*, 617–623. [CrossRef]

87. Brian, P.; Curtis, P.; Grove, J.; Hemming, H.; McGowan, J. Gladiolic acid: An antifungal and antibacterial metabolic product of *Penicillium gladioli* McCull and Thom. *Nature* **1946**, *157*, 697. [CrossRef]

88. Suzuki, K.; Nozawa, K.; Nakajima, S.; Udagawa, S.I.; Kawai, K.I. Isolation and structures of antibacterial binaphtho-α-pyrones, talaroderxines A and B, from *Talaromyces derxii*. *Chem. Pharm. Bull.* **1992**, *40*, 1116–1119. [CrossRef] [PubMed]

89. Zhao, Q.H.; Yang, Z.D.; Shu, Z.M.; Wang, Y.G.; Wang, M.G. Secondary metabolites and biological activities of *Talaromyces* sp. LGT-2, an endophytic fungus from *Tripterygium wilfordii*. *Iran. J. Pharm. Res.* **2016**, *15*, 453–457. [PubMed]

90. Houbraken, J.; Samson, R.A. Phylogeny of *Penicillium* and the segregation of Trichocomaceae into three families. *Stud. Mycol.* **2011**, *70*, 1–51. [CrossRef] [PubMed]

91. Jin, P.; Zuo, W.; Guo, Z.; Mei, W.; Dai, H. Metabolites from the endophytic fungus *Penicillium* sp. FJ-1 of Ceriops tagal. *Acta Pharm. Sin.* **2013**, *48*, 1688–1691.

92. Liu, F.; Cai, X.L.; Yang, H.; Xia, X.K.; Guo, Z.Y.; Yuan, J.; Li, M.F.; She, Z.G.; Lin, Y.C. The bioactive metabolites of the mangrove endophytic fungus *Talaromyces* sp. ZH-154 isolated from *Kandelia candel* (L.) Druce. *Planta Med.* **2010**, *76*, 185–189. [CrossRef] [PubMed]

93. Wu, B.; Ohlendorf, B.; Oesker, V.; Wiese, J.; Malien, S.; Schmaljohann, R.; Imhoff, J.F. Acetylcholinesterase inhibitors from a marine fungus *Talaromyces* sp. strain LF458. *Mar. Biotechnol.* **2015**, *17*, 110–119. [CrossRef]

94. Stark, A.A.; Townsend, J.M.; Wogan, G.N.; Demain, A.L.; Manmade, A.; Ghosh, A.C. Mutagenicity and antibacterial activity of mycotoxins produced by *Penicillium islandicum* Sopp and *Penicillium rugulosum*. *J. Environ. Pathol. Toxicol.* **1978**, *2*, 313–324.

95. Brian, P.; Curtis, P.; Hemming, H.; Norris, G. Wortmannin, an antibiotic produced by *Penicillium wortmanni*. *Trans. Br. Mycol. Soc.* **1957**, *40*, 365–368. [CrossRef]

96. Bara, R.; Zerfass, I.; Aly, A.H.; Goldbach-Gecke, H.; Raghavan, V.; Sass, P.; Mandi, A.; Wray, V.; Polavarapu, P.L.; Pretsch, A. Atropisomeric dihydroanthracenones as inhibitors of multiresistant *Staphylococcus aureus*. *J. Med. Chem.* **2013**, *56*, 3257–3272. [CrossRef]

97. Miao, F.; Yang, R.; Chen, D.D.; Wang, Y.; Qin, B.F.; Yang, X.J.; Zhou, L. Isolation, identification and antimicrobial activities of two secondary metabolites of *Talaromyces verruculosus*. *Molecules* **2012**, *17*, 14091–14098. [CrossRef] [PubMed]

98. Chen, S.; Liu, Y.; Liu, Z.; Cai, R.; Lu, Y.; Huang, X.; She, Z. Isocoumarins and benzofurans from the mangrove endophytic fungus *Talaromyces amestolkiae* possess α-glucosidase inhibitory and antibacterial activities. *RSC Adv.* **2016**, *6*, 26412–26420. [CrossRef]

99. Yamazaki, M.; Okuyama, E. Isolation and structures of oxaphenalenone dimers from *Talaromyces bacillosporus*. *Chem. Pharm. Bull.* **1980**, *28*, 3649–3655. [CrossRef]

100. Zhai, M.M.; Niu, H.T.; Li, J.; Xiao, H.; Shi, Y.P.; Di, D.L.; Crews, P.; Wu, Q.X. Talaromycolides A–C, novel phenyl-substituted phthalides isolated from the green Chinese onion-derived fungus *Talaromyces pinophilus* AF-02. *J. Agric. Food Chem.* **2015**, *63*, 9558–9564. [CrossRef]

101. Ren, J.; Ding, S.S.; Zhu, A.; Cao, F.; Zhu, H.J. Bioactive azaphilone derivatives from the fungus *Talaromyces aculeatus*. *J. Nat. Prod.* **2017**, *80*, 2199–2203. [CrossRef]

102. Huang, H.; Liu, T.; Wu, X.; Guo, J.; Lan, X.; Zhu, Q.; Zheng, X.; Zhang, K. A new antibacterial chromone derivative from mangrove-derived fungus *Penicillium aculeatum* (No. 9EB). *Nat. Prod. Res.* **2017**, *31*, 2593–2598. [CrossRef]

103. Li, H.L.; Li, X.M.; Liu, H.; Meng, L.H.; Wang, B.G. Two new diphenylketones and a new xanthone from *Talaromyces islandicus* EN-501, an endophytic fungus derived from the marine red alga *Laurencia okamurai*. *Mar. Drugs* **2016**, *14*, 223. [CrossRef]

104. Cai, R.; Chen, S.; Long, Y.; Li, C.; Huang, X.; She, Z. Depsidones from *Talaromyces stipitatus* SK-4, an endophytic fungus of the mangrove plant *Acanthus ilicifolius*. *Phytochem. Lett.* **2017**, *20*, 196–199. [CrossRef]

105. Li, G.H.; Zheng, L.J.; Liu, F.F.; Dang, L.Z.; Li, L.; Huang, R.; Zhang, K.Q. New cyclopentenones from strain *Trichoderma* sp. YLF-3. *Nat. Prod. Res.* **2009**, *23*, 1431–1435. [CrossRef]

106. Adachi, T.; Aoki, H.; Osawa, T.; Namiki, M.; Yamane, T.; Ashida, T. Structure of trichodermaol, antibacterial substance produced in combined culture of *Trichoderma* sp. with Fusarium oxysporum or *Fusarium solani*. *Chem. Lett.* **1983**, *12*, 923–926. [CrossRef]

107. Zhang, J.C.; Chen, G.Y.; Li, X.Z.; Hu, M.; Wang, B.Y.; Ruan, B.H.; Zhou, H.; Zhao, L.X.; Zhou, J.; Ding, Z.T. Phytotoxic, antibacterial, and antioxidant activities of mycotoxins and other metabolites from *Trichoderma* sp. *Nat. Prod. Res.* **2017**, *31*, 2745–2752. [CrossRef] [PubMed]

108. Rebuffat, S.; Goulard, C.; Hlimi, S.; Bodo, B. Two unprecedented natural Aib-peptides with the (Xaa-Yaa-Aib-Pro) motif and an unusual C-terminus: Structures, membrane-modifying and antibacterial properties of pseudokonins KL III and KL VI from the fungus *Trichoderma pseudokoningii*. *J. Pept. Sci.* **2000**, *6*, 519–533. [CrossRef]

109. Ooka, T.; Shimojima, Y.; Akimoto, T.; Takeda, I.; Senoh, S.; Abe, J. A new antibacterial peptide "suzukacillin". *Agric. Biol. Chem.* **1966**, *30*, 700–702. [CrossRef]

110. Brian, P.; Curtis, P.; Hemming, H.; McGowan, J. The production of viridin by pigment-forming strains of *Trichoderma viride*. *Ann. Appl. Biol.* **1946**, *33*, 190–200. [CrossRef] [PubMed]

111. Berg, A.; Kemami Wangun, H.V.; Nkengfack, A.E.; Schlegel, B. Lignoren, a new sesquiterpenoid metabolite from *Trichoderma lignorum* HKI 0257. *J. Basic Microbiol.* **2004**, *44*, 317–319. [CrossRef] [PubMed]

112. Reino, J.L.; Guerrero, R.F.; Hernández-Galán, R.; Collado, I.G. Secondary metabolites from species of the biocontrol agent *Trichoderma*. *Phytochem. Rev.* **2008**, *7*, 89–123. [CrossRef]

113. Tarus, P.K.; Lang'at-Thoruwa, C.C.; Wanyonyi, A.W.; Chhabra, S.C. Bioactive metabolites from *Trichoderma harzianum* and *Trichoderma longibrachiatum*. *Bull. Chem. Soc. Ethiop.* **2003**, *17*, 185–190.

114. Dang, L.; Li, G.; Yang, Z.; Luo, S.; Zheng, X.; Zhang, K. Chemical constituents from the endophytic fungus *Trichoderma ovalisporum* isolated from *Panax notoginseng*. *Ann. Microbiol.* **2010**, *60*, 317–320. [CrossRef]

115. Khamthong, N.; Rukachaisirikul, V.; Tadpetch, K.; Kaewpet, M.; Phongpaichit, S.; Preedanon, S.; Sakayaroj, J. Tetrahydroanthraquinone and xanthone derivatives from the marine-derived fungus *Trichoderma aureoviride* PSU-F95. *Arch. Pharmacal Res.* **2012**, *35*, 461–468. [CrossRef]

116. Gardes, M.; Bruns, T.D. ITS primers with enhanced specificity for basidiomycetes-application to the identification of mycorrhizae and rusts. *Mol. Ecol.* **1993**, *2*, 113–118. [CrossRef]

117. Bellemain, E.; Carlsen, T.; Brochmann, C.; Coissac, E.; Taberlet, P.; Kauserud, H. ITS as an environmental DNA barcode for fungi: An in silico approach reveals potential PCR biases. *BMC Microbiol.* **2010**, *10*, 189. [CrossRef] [PubMed]

118. Glass, N.L.; Donaldson, G.C. Development of primer sets designed for use with the PCR to amplify conserved genes from filamentous ascomycetes. *Appl. Environ. Microbiol.* **1995**, *61*, 1323–1330. [PubMed]

119. O'Donnell, K.; Kistler, H.C.; Cigelnik, E.; Ploetz, R.C. Multiple evolutionary origins of the fungus causing Panama disease of banana: Concordant evidence from nuclear and mitochondrial gene genealogies. *Proc. Natl. Acad. Sci. USA* **1998**, *95*, 2044–2049. [CrossRef] [PubMed]

120. Kumar, S.; Stecher, G.; Tamura, K. MEGA7: Molecular evolutionary genetics analysis version 7.0 for bigger datasets. *Mol. Biol. Evol.* **2016**, *33*, 1870–1874. [CrossRef]

121. Katoh, K.; Standley, D.M. MAFFT multiple sequence alignment software version 7: Improvements in performance and usability. *Mol. Biol. Evol.* **2013**, *30*, 772–780. [CrossRef]

122. Maddison, D.R.; Maddison, W.P. *MacClade 4: Analysis of Phylogeny and Character Evolution*; Version 4.08 a. Sinauer Associates: Sunderland, MA, USA, 2005.

123. Nylander, J.A.A. *MrModeltest v2. Program distributed by the author*; Evolutionary Biology Center, Uppsala University: Uppsala, Sweden, 2004.

124. Ronquist, F.; Huelsenbeck, J.P. MrBayes 3: Bayesian phylogenetic inference under mixed models. *Bioinformatics* **2003**, *19*, 1572–1574. [CrossRef]

125. Ihrmark, K.; Bödeker, I.; Cruz-Martinez, K.; Friberg, H.; Kubartova, A.; Schenck, J.; Strid, Y.; Stenlid, J.; Brandström-Durling, M.; Clemmensen, K.E.; et al. New primers to amplify the fungal ITS2 region—Evaluation by 454-sequencing of artificial and natural communities. *FEMS Microbiol. Ecol.* **2012**, *82*, 666–677. [CrossRef]

126. Caporaso, J.G.; Kuczynski, J.; Stombaugh, J.; Bittinger, K.; Bushman, F.D.; Costello, E.K.; Fierer, N.; Pena, A.G.; Goodrich, J.K.; Gordon, J.I.; et al. QIIME allows analysis of high-throughput community sequencing data. *Nat. Methods* **2010**, *7*, 335–336. [CrossRef]

127. Rognes, T.; Flouri, T.; Nichols, B.; Quince, C.; Mahé, F. VSEARCH: A versatile open source tool for metagenomics. *PeerJ* **2016**, *4*, e2584. [CrossRef]

128. Nilsson, R.H.; Tedersoo, L.; Ryberg, M.; Kristiansson, E.; Hartmann, M.; Unterseher, M.; Porter, T.M.; Bengtsson-Palme, J.; Walker, D.M.; de Sousa, F.; et al. A comprehensive, automatically updated fungal ITS sequence dataset for reference-based chimera control in environmental sequencing efforts. *Microbes Environ.* **2015**, *30*, 145–150. [CrossRef]

129. Segata, N.; Izard, J.; Waldron, L.; Gevers, D.; Miropolsky, L.; Garrett, W.S.; Huttenhower, C. Metagenomic biomarker discovery and explanation. *Genome Biol.* **2011**, *12*, R60. [CrossRef] [PubMed]

130. Dufrêne, M.; Legendre, P. Species assemblages and indicator species: The need for a flexible asymmetrical approach. *Ecol. Monogr.* **1997**, *67*, 345–366. [CrossRef]

131. Wang, Y.; Naumann, U.; Wright, S.T.; Warton, D.I. Mvabund—An R package for model-based analysis of multivariate abundance data. *Methods Ecol. Evol.* **2012**, *3*, 471–474. [CrossRef]

132. Heo, Y.M.; Kim, K.; Ryu, S.M.; Kwon, S.L.; Park, M.Y.; Kang, J.E.; Hong, J.H.; Lim, Y.W.; Kim, C.; Kim, B.S.; et al. Diversity and ecology of marine algicolous *Arthrinium* species as a source of bioactive natural products. *Mar. Drugs* **2018**, *16*, 508. [CrossRef]

133. Lai, H.Y.; Lim, Y.Y.; Tan, S.P. Antioxidative, tyrosinase inhibiting and antibacterial activities of leaf extracts from medicinal ferns. *Biosci. Biotechnol. Biochem.* **2009**, *73*, 1362–1366. [CrossRef]

134. Kim, J.D.; Han, J.W.; Hwang, I.C.; Lee, D.; Kim, B.S. Identification and biocontrol efficacy of *Streptomyces miharaensis* producing filipin III against *Fusarium wilt*. *J. Basic Microbiol.* **2012**, *52*, 150–159. [CrossRef]

135. Kang, J.E.; Han, J.W.; Jeon, B.J.; Kim, B.S. Efficacies of quorum sensing inhibitors, piericidin A and glucopiericidin A, produced by *Streptomyces xanthocidicus* KPP01532 for the control of potato soft rot caused by *Erwinia carotovora* subsp. *atroseptica*. *Microbiol. Res.* **2016**, *184*, 32–41. [CrossRef]

Article

Discovery of Bioactive Indole-Diketopiperazines from the Marine-Derived Fungus *Penicillium brasilianum* Aided by Genomic Information

Ya-Hui Zhang [1,2,†], Ce Geng [3,†], Xing-Wang Zhang [3], Hua-Jie Zhu [2], Chang-Lun Shao [1,4], Fei Cao [2,*] and Chang-Yun Wang [1,4,5,*]

[1] Key Laboratory of Marine Drugs, the Ministry of Education of China, School of Medicine and Pharmacy, Ocean University of China, Qingdao 266003, China
[2] College of Pharmaceutical Sciences, Key Laboratory of Pharmaceutical Quality Control of Hebei Province, Hebei University, Baoding 071002, China
[3] Shandong Provincial Key Laboratory of Synthetic Biology, CAS Key Laboratory of Biofuels at Qingdao Institute of Bioenergy and Bioprocess Technology, Chinese Academy of Sciences, Qingdao 266101, China
[4] Laboratory for Marine Drugs and Bioproducts, Qingdao National Laboratory for Marine Science and Technology, Qingdao 266237, China
[5] Institute of Evolution & Marine Biodiversity, Ocean University of China, Qingdao 266003, China
* Correspondence: caofei542927001@163.com (F.C.); changyun@ouc.edu.cn (C.-Y.W.)
† These authors contributed equally to this work.

Received: 2 August 2019; Accepted: 29 August 2019; Published: 1 September 2019

Abstract: Identification and analysis of the whole genome of the marine-derived fungus *Penicillium brasilianum* HBU-136 revealed the presence of an interesting biosynthetic gene cluster (BGC) for non-ribosomal peptide synthetases (NRPS), highly homologous to the BGCs of indole-diketopiperazine derivatives. With the aid of genomic analysis, eight indole-diketopiperazines (**1–8**), including three new compounds, spirotryprostatin G (**1**), and cyclotryprostatins F and G (**2** and **3**), were obtained by large-scale cultivation of the fungal strain HBU-136 using rice medium with 1.0% $MgCl_2$. The absolute configurations of **1–3** were determined by comparison of their experimental electronic circular dichroism (ECD) with calculated ECD spectra. Selective cytotoxicities were observed for compounds **1** and **4** against HL-60 cell line with the IC_{50} values of 6.0 and 7.9 μM, respectively, whereas **2**, **3**, and **5** against MCF-7 cell line with the IC_{50} values of 7.6, 10.8, and 5.1 μM, respectively.

Keywords: biosynthetic gene cluster (BGC); indole-diketopiperazine; *Penicillium brasilianum*; cytotoxicities

1. Introduction

In the past decades, marine-derived fungi have attracted more and more attention to medicinal chemists because of their ability to produce biologically active compounds with diverse structures for drug discovery [1]. Especially, one fungus species could generate structurally versatile natural products because it consists of a large number of biosynthetic gene clusters (BGCs) associated with its secondary metabolism [2,3]. For example, several bioactive compounds have been generated by the same fungus from different sources, including antifungal diketopiperazine-type alkaloids [4], antibacterial dimeric diketopiperazines [5], cytotoxic prenyl asteltoxin derivatives [6], and anti-virus highly oxygenated cyclopiazonic acid-derived alkaloids [7]. However, it has been proven that most of the genes encoding secondary metabolites in fungi are cryptic under certain culture conditions [8]. In recent years, due to the advances in genome sequencing and bioinformatics-based predictions of encoded natural products in BGCs, it has been enabled to obtain structurally-unique secondary metabolites from fungi with the assistance of genomic information [8–10].

During our investigation focused on discovering bioactive natural products from marine-derived fungi [11–14], two known cytotoxic diketopiperazine alkaloids, spirocyclic diketopiperazine alkaloid (**4**) [15] and cyclotryprostatin B (**5**) [16], were discovered from the marine-derived fungus *Penicillium brasilianum* HBU-136 by using rice medium in the first large-scaled fermentation. After gaining these two known diketopiperazine alkaloid derivatives, we tried to locate their BGC in the genome of the fungus HBU-136. A search of the genome sequence using antiSMASH 4.2 found that the strain harbors multiple BGCs for non-ribosomal peptide synthetases (NRPS), one of which shows high similarity to the BGC of fumitremorgins A-C in the fungus *Aspergillus fumigatus* [17–19]. By gene-to-gene comparisons, it is disclosed that the main difference between the BGC in HBU-136 (*ctp*) against in *Aspergillus fumigatus* (*ftm*) lies at the tailoring genes (Table S2 and Figure S45), which suggested that the BGC in HBU-136 may having the ability to produce new derivatives. Based on this information, we tried more fermentation conditions with LC-MS analysis, and finally found three new indole-diketopiperazines (**1–3**) from the rice medium with 1.0% $MgCl_2$ fermentation. Herein, we report the isolation, structural characterization, putative biosynthetic relation and biological activities of these compounds.

2. Results and Discussion

2.1. Identification and Analysis of the Indole-Diketopiperazine BGC

The whole genome sequence of *Penicillium brasilianum* HBU-136 was sequenced within HiSeq X10 platform (Illumina, CA, USA) (150 fold) and further assembled into 937 contigs (35.5 Mb) with *de novo* assembly software ABySS and SPAdes. For further accurate BGCs mining and comparison, all the genes were predicted by gene finding software GeneMarks ES/ET and Augustus with self-training. Then, 50 secondary metabolite gene clusters were proposed by using antiSMASH 4.2 as a gene cluster mining tool (Table S1). Thereinto, 12 out of the 50 BGCs were described as NRPS, including a BGC highly homologous to the fumitremorgin BGC (8 out of 9 genes show similarity Table S2) [17–19]. This highly similar gene cluster involved an NRPS gene (*ctpNRPS*), three cytochrome P450 genes (*ctpP450-1/2/3*), one oxmethyltransferase gene (*ctpOMT*), two prenyltransferase genes (*ctpPT-1/2*), and one putative oxidase gene (*ctpOx*) (Table S2).

By gene-to-gene comparison, it suggested that the proposed gene clusters were consistent well with the assembly of the dipeptide skeleton for indole-diketopiperazines (Figure 1). The key biosynthetic step of dipeptide skeleton started from two successive amidation reactions catalyzed by the *ctpNRPS* gene and led to the formation of brevianamide F [20,21]. Then *ctpPT-1* gene mediated the isopentane addition and delivered the product directly to *ctpP450-2* gene, who catalyzed the tetrahydropyridine ring formation, or to *ctpOMT* gene to get the methoxyl first before the ring-formation step [20,21]. The formed cyclotryprostatin skeleton could be further transformed into spirotryprostatin-type by *ctpP450-3* gene via a radical formation and migration eliciting semipinacol-type rearrangement mechanism [22]. P450s, the most versatile biocatalysts in nature, have been proved to catalyze such uncommon reactions during skeleton constructing processes in many cases [23]. After the core skeletons constructed, the structural diversity of both spirotryprostatins and cyclotryprostatins mainly come from the tailoring reactions of *ctpP450-3* gene to complete hydroxylation reactions, *ctpP450-1* gene to catalyze the benzene hydroxylation, and *ctpOMT* gene to mediate the methylation reaction.

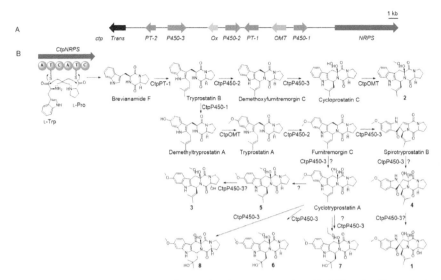

Figure 1. (**A**) The biosynthetic gene cluster of cyclotryprostatin from *Penicillium brasilianum* (*ctp*). (**B**) Proposed biosynthetic pathway of the isolated indole-diketopiperazines.

More intriguingly, the gene *ctp*P450-3, whose corresponding gene in *ftm* is the multi-functional gene *ftmG*, was also inferred to encode a muti-functional P450 according to their sequence identity, suggesting that new polyhydroxylated indole-diketopiperazines could be generated by the fungal strain HBU-136. At this point, we tried eight different media conditions (C1–C8) with LC-MS screening to mine more biosynthetic products for *ctp*. Guided by HPLC-MS, it was found that three previously unrecognized compounds exhibiting the typical indole UV spectrum were produced by the strain HBU-136 when 1.0% $MgCl_2$ was added to rice medium (C-8). According to previous literature, the addition of $MgCl_2$ into the medium was reported to impact the secondary metabolism [13,24], which was confirmed once again by our research. Finally, eight indole-diketopiperazines (1–8), including three new compounds, spirotryprostatin L (**1**) and cyclotryprostatins F–G (**2–3**), and five known analogs, spirocyclic diketopiperazine alkaloid (**4**) [15], cyclotryprostatin B (**5**) [16], 20-hydroxycyclotryprostatin B (**6**) [25], 12β-hydroxy-13α-ethoxyverruculogen TR-2 (**7**) [26], and an unnamed compound (**8**) [27], were then obtained (Figure 2).

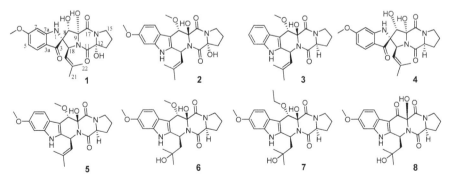

Figure 2. Chemical structures of compounds 1–8.

2.2. Structure Elucidation

Spirotryprostatin L (**1**), a yellow amorphous powder, was obtained with the molecular formula $C_{22}H_{25}N_3O_7$ (12 degrees of unsaturation) by positive HRESIMS. The IR absorption bands at 3447 cm^{-1} and 1647 cm^{-1} indicated the presences of hydroxyl and carbonyl groups. The characteristic UV absorption maxima at 220, 249, 287, 387 nm suggested a spirotryprostatin skeleton for **1** [15,16]. The NMR signals in **1** (Tables 1 and 2) were representative of one keto carbonyl (C-3), two amide carbonyls (C-11 and C-17), one 1,3,4-trisubstituted aromatic unit (C-3a, C-4, C-5, C-6, C-7, and C-7a), and three methyls (one oxygenated) (6-OCH₃, C-22, and C-21). These structural features suggested a prenylated indole-diketopiperazine nucleus of **1**, structurally similar to the known compound **4** which was previously obtained from the fungus *Aspergillus fumigatus* [15]. A detailed analysis for HMBC correlations of **1** (from 12-OH to C-12 and C-13) (Figure 3) indicated that an additional hydroxyl group [δ_H 5.14 (1H, brs, 12-OH) and δ_C 90.2 (C-12)] was attached to C-12. Based on the systematic 2D NMR data analyses, the plane structure of **1** was determined.

Table 1. ^{1}H NMR Data (δ) of **1–3** (500 MHz, CDCl₃, *J* in Hz).

No.	1	2	3
1		7.86, s	7.97, s
4	7.50, d (8.7)	7.48, d (8.4)	7.58, d (7.6)
5	6.45, dd (8.7, 2.0)	6.80, dd (8.4, 2.0)	7.17, dtd (8.0, 7.6, 0.9)
6			7.17, dtd (8.0, 7.6, 0.9)
7	6.27, d (2.0)	6.86, d (2.0)	7.37, d (7.6)
8	4.70, s	5.02, s	4.79, s
12			4.38, dd (10.7, 6.3)
13	2.35, m	2.40, m	2.50, m
	2.27, m	1.98, m	2.01, m
14	2.15, m	2.10, s	2.12, m
	2.00, m	1.98, m	2.01, m
15	3.74, m	3.80, m	3.76, m
	3.54, m	3.70, m	3.71, m
18	4.83, d (9.5)	6.36, d (9.6)	6.68, d (9.7)
19	4.79, d (9.5)	5.33, d (9.6)	5.57, d (9.7)
21	1.56, s	2.04, s	1.80, s
22	1.79, s	1.77, s	2.07, s
8-OH	4.96, brs		
12-OH	5.14, brs		
6-OCH₃	3.85, s	3.82, s	
9-OH	8.51, brs		
8-OCH₃		3.30, s	3.38, s

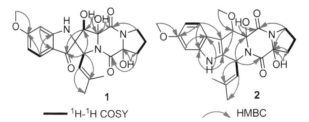

1   **2**

—— ^{1}H-^{1}H COSY   HMBC

Figure 3. ^{1}H-^{1}H COSY and key HMBC correlations of **1** and **2**. In order to assign the relative configuration of **1**, its NOESY experiment was carried out.

Table 2. ^{13}C NMR Data (δ) of **1–3** (125 MHz, CDCl$_3$).

No.	1	2	3
2	75.0, C	132.7, C	135.2, C
3	200.6, C	104.8, C	105.7, C
3a	112.4, C	122.5, C	128.5, C
4	127.4, CH	118.9, CH	118.2, CH
5	110.8, CH	110.1, CH	120.7, CH
6	169.7, C	156.6, C	122.3, CH
7	94.9, CH	95.3, CH	111.3, CH
7a	165.4, C	137.1, C	135.9, C
8	73.9, CH	74.9, CH	76.8, CH
9	86.1, C	85.7, C	84.9, C
11	167.4, C	166.3, C	167.2, C
12	90.2, C	86.7, C	60.1, CH
13	35.1, CH$_2$	36.5, CH$_2$	29.8, CH$_2$
14	21.3, CH$_2$	19.2, CH$_2$	22.3, CH$_2$
15	45.3, CH$_2$	45.4, CH$_2$	46.0, CH$_2$
17	166.0, C	165.9, C	166.0, C
18	55.7, CH	49.3, CH	49.2, CH
19	119.4, CH	122.8, CH	123.6, CH
20	142.4, C	138.2, C	138.2, C
21	26.1, CH$_3$	18.4, CH$_3$	26.2, CH$_3$
22	18.8, CH$_3$	26.1, CH$_3$	18.4, CH$_3$
6-OCH$_3$	56.0, CH$_3$	55.8, OCH$_3$	
8-OCH$_3$		57.0, OCH$_3$	56.8, OCH$_3$

The NOESY correlations of **1** between H-7 and 8-OH, 8-OH and 9-OH, 9-OH and 12-OH, 12-OH and H-18, and H-8 and H-19 (Figure 4) suggested that H-18, 8-OH, 9-OH, and 12-OH were located on the same face of the molecule, and H-7 had a *cis* relationship with 8-OH/9-OH/12-OH/H-18. The electronic circular dichroism (ECD) spectrum of **1** was tested and calculated to investigate the absolute configuration of **1**. Time-dependent density functional theory (TD-DFT) method was used for ECD calculation of the molecule (2*S*,8*S*,9*R*,12*R*,18*S*)-**1** at the B3LYP/6-311++G(2d,p)//B3LYP/6-311+G(d) level. Boltzmann statistics with a standard deviation of σ 0.3 eV was used for the ECD simulation. The calculated ECD spectrum of (2*S*,8*S*,9*R*,12*R*,18*S*)-**1** was in good accordance with the experimental ECD data of **1** (Figure 5), revealing that the absolute configuration of **1** should be 2*S*,8*S*,9*R*,12*R*,18*S*. Additionally, the above conclusion was verified by the fact that the experimental ECD spectrum and the specific optical rotation (OR) value of **1** were similar to those of **4** ([*a*]$_D^{20}$ = + 144.6 (*c* 1.00, MeOH) for **1** vs. [*a*]$_D^{20}$ = + 147.2 (*c* 0.1, CHCl$_3$) for **4**) [15]. Thus, compound **1** was defined as a 12-hydroxylation derivative of **4**.

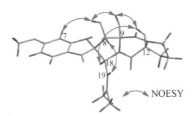

Figure 4. The key NOESY correlations of **1**.

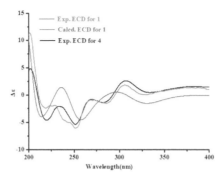

Figure 5. Experimental ECD of **1** and **4**, and calculated ECD of (2*S*,8*S*,9*R*,12*R*,18*S*)-**1**.

Cyclotryprostatin F (**2**) was also obtained as a yellow amorphous powder, whose molecular formula was deduced as $C_{23}H_{27}N_3O_6$ from HRESIMS. The NMR data of **2** (Table 1; Table 2) suggested that it also possessed a diketopiperazine skeleton, which was considered similar to the known compound **5** [16] through detailed NMR analyses of them. Combining the upfield shifted C-11 (δ_C 166.3 in **2**; δ_C 167.0 in **5**) and C-14 (δ_C 19.2 in **2**; δ_C 22.1 in **5**), and downfield shifted C-12 (δ_C 86.7 in **2**; δ_C 59.9 in **5**) and C-13 (δ_C 36.5 in **2**; δ_C 29.7 in **5**) of **2** with the change of the molecular weight from **5** to **2** suggested that **2** was a 12-hydroxylation derivative of **5**. This deduction could be confirmed by the key HMBC correlations from H-13 and H-15 to C-12 in **2** (Figure 3). Similar ^1H-^1H coupling constants, NOESY correlations and ECD Cotton effects (Figure 6) of **2** and **5** implied that **2** had the 8*S*,9*S*,12*R*,18*S* absolute configuration, which was also confirmed by TD-DFT ECD calculation of (8*S*,9*S*,12*R*,18*S*)-**2**.

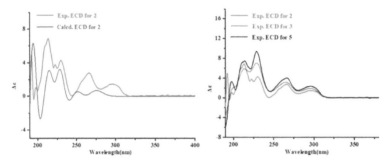

Figure 6. Experimental ECD of **2**, **3** and **5**, and calculated ECD of (8*S*,9*S*,12*R*,18*S*)-**2**.

Cyclotryprostatin G (**3**) was also isolated as an analog of **5**, deduced from the similar NMR data of them. Compound **3** was the demethylation derivative of **5** at C-6, which was verified by ^1H-^1H COSY (H-4/H-5/H-6/H-7) and HMBC (H-6 and C-4, and H-6 and C-7a) experiments for **3**. The stereochemistry of **3** was suggested to be 8*S*,9*S*,12*S*,18*S* by comparing the NOESY and ECD spectra of **3** to those of **5** (Figure 6).

It is deduced that, the newborn OH-12 in **1** and **3** may be catalyzed by the multi-functional P450 *ctpP450-3* (Figure 1). And the new formed OMe-8 in **23** may be introduced by the methyltransferase encoded gene *ctpOMT* or other genes outside the reported boundary, or they were formed during our isolation processes.

2.3. Biological Activities Screening

It has been reported that many indole-diketopiperazine derivatives displayed a wide range of biological effects, such as cytotoxic [16,28] andantibacterial activities [29]. In the present study, all of the isolated compounds (**1–8**) were tested for their cytotoxic and antibacterial activities. Among them, **1** and **4** displayed selective cytotoxicities against HL-60 cell line with the IC_{50} values of 6.0 and 7.9 μM, respectively. Whereas compounds **2**, **3**, and **5** exhibited activities against MCF-7 cell line with the IC_{50} values of 7.6, 10.8, and 5.1 μM, respectively. However, all of the metabolites appeared to be inactive in antibacterial and antifungal assays (MIC > 25 μM).

3. Materials and Methods

3.1. Instrumentation

Optical rotations were measured on a JASCO P-1020 digital polarimeter (Jasco Corp., Tokyo, Japan). UV spectra were recorded on a Thermo Scientific Multiskan GO microplate spectrophotometer (Thermo Scientific Co., Waltham, MA, USA) in MeOH. ECD spectra were recorded on a JASCO J-815 circular dichroism spectrometer (JASCO Electric Co., Ltd., Tokyo, Japan). IR spectra were determined on a Nicolet-Nexus-470 spectrometer (Thermo Electron Co., Madison, WI, USA) using KBr pellets. 1D and 2D NMR spectroscopic data (TMS as an internal standard) were acquired on Agilent DD2-500 (JEOL, Tokyo, Japan) and Bruker AV-600 NMR spectrometer (Bruker BioSpin GmbH Co., Rheinstetten, Germany). ESIMS and HRESIMS spectra were obtained from a Micromass Q-TOF spectrometer (Waters Corp., Manchester, UK), and a Thermo Scientific LTQ Orbitrap XL spectrometer (Thermo Fisher Scientific, Bremen, Germany). Semipreparative HPLC was performed on a Shimadzu LC-20 AT system using a RP-C18 (Waters, 5 μm, 10 × 250 mm) column.

3.2. Genome Sequencing and Bioinformatics Analysis

The fungus *Penicillium brasilianum* HBU-136 was cultured in PDB medium for 30 h at 25 °C, then high-quality genomic DNA was isolated from cultured cells using Fungal DNA Kit (OMEGA Bio-Tek, E.Z.N.A.®, (Norcross, GA, USA). All the purified DNA was used to construct 350 bp DNA library with DNA library preparation kit (Illumina, CA, USA), and was sequenced on HiSeq X10 instrument, employing paired-end 150 base reads (Mega genomics, CN). *De novo* assembling of the draft genome was performed by using the software ABySS 2.15 (Vancouver, NA, Canada) and SPAdes 3.11 (St. Petersburg, EU, Russia), with multiple k-mers specified as "-k 33,55,77,99,117" and coverage fold of 160X PE clean reads. All the assemble scaffold sequences were tested by quality assessment tool QUAST 4.0, the last draft genome sequence was filtered by over 500 bp and was determined via the largest N50 number. Gene functions were proposed by the online BLAST program (http://blast.ncbi.nlm.nih.gov/) and Conserved Domain Database (CDD) at the NCBI server (https://www.ncbi.nlm.nih.gov/). This Whole Genome Shotgun project of the fungus HBU-136 has been deposited at DDBJ/ENA/GenBank under the accession SSWP00000000. The version described in this paper is version SSWP01000000.

3.3. Fungal Material

The fungus *Penicillium brasilianum* HBU-136 was collected from the Bohai Sea (Huanghua, Hebei Province, China, June 2016). The strain with the NCBI GenBank accession number MH377073 was deposited in the College of Pharmaceutical Sciences, Hebei University, China.

3.4. Fermentation and Purification

The first fermentation was carried out for the fungus HBU-136 using rice medium (80 mL water and 80 g rice in 1 L Erlenmeyer flasks, 40 flasks) at 28 °C for 45 days. After cultivation, the fermented rice substrate was extracted with the mixture of CH_2Cl_2/MeOH (1:1, 500 mL for each flask) for five times and EtOAc for two times successively. The two extract solvents were combined as their TLC profiles were almost the same. The combined extract was concentrated to give a residue (30 g), which was further subjected to silica gel column chromatography (CC), eluting with an EtOAc−petroleum ether stepped gradient elution (0%, 20%, 40%, 60%, 80%, 100%), to offer six fractions Fr.1−Fr.6. Fr.5 was separated by twice Sephadex LH-20 CC [petroleum ether−MeOH−$CHCl_3$ (v/v, 2:1:1)] to afford four subfractions Fr.5-1–Fr.5-4. Then, Fr.5-3 was fractionated by Sephadex LH-20 CC [MeOH–CHCl$_3$ (v/v, 1:1)] and silica gel CC (stepped gradient, EtOAc–petroleum ether), and further purified by semi-preparative HPLC (MeOH–H$_2$O, 50:50, 2 mL/min) to afford **4** (3.1 mg) and **5** (6.7 mg).

The second large-scale fermentation (100 flasks) for the fungus HBU-136 was performed for 45 days in a modified rice medium (including 80 mL water, 80 g rice, and 0.8 g MgCl$_2$ in 1 L Erlenmeyer flask). The extract was obtained and purified by the same approaches as first fermentation. Compounds **1** (1.8 mg), **4** (5.2 mg), **5** (4.8 mg), **7** (1.5 mg), and **8** (6.2 mg) were given from fraction Fr.5. Compounds **2** (4.0 mg), **3** (3.2 mg), and **6** (4.3 mg) were isolated from fraction Fr.6.

Spirotryprostatin G (**1**): Yellow amorphous powder; $[\alpha]_D^{20}$ = + 44.6 (*c* 1.00, MeOH); UV (MeOH) λ_{max} (log ε) 204 (4.05), 220 (4.02), 249 (3.96), 287 (3.77), 387 (3.17) nm; CD (0.75 mM, MeOH) λ_{max} ($\Delta\varepsilon$) 201 (+20.7), 219 (−4.3), 239 (−3.4), 252 (−10.9), 327 (+0.16) nm; IR (KBr) v_{max} 3447, 2360 and 1647 cm^{-1}; ^{1}H and ^{13}C NMR, see Table 1; Table 2; HRESIMS *m/z* 444.1757 [M + H]$^+$ (calcd for C$_{22}$H$_{26}$N$_3$O$_7$, 444.1765 [M + Na]$^+$), 466.1578 [M + Na]$^+$ (calcd for C$_{22}$H$_{25}$N$_3$O$_7$Na, 466.1585 [M + Na]$^+$).

Cyclotryprostatin F (**2**): Yellow amorphous powder; $[\alpha]_D^{20}$ = + 108 (*c* 1.00, MeOH); UV (MeOH) λ_{max} (log ε) 220 (4.30), 290 (3.70) nm; CD (0.75 mM, MeOH) λ_{max} ($\Delta\varepsilon$) 213 (+13.8), 230 (+8.6), 266 (+5.6), 319 (−0.01) nm; ^{1}H and ^{13}C NMR, see Table 1; Table 2; HRESIMS *m/z* 440.1837 [M − H]$^-$ (calcd for C$_{23}$H$_{26}$N$_3$O$_6$, 440.1827 [M − H]$^-$).

Cyclotryprostatin G (**3**): Yellow amorphous powder; $[\alpha]_D^{20}$ = + 101 (*c* 0.95, MeOH); UV (MeOH) λ_{max} (log ε) 223 (4.45), 272 (3.93) nm; CD (1.05 mM, MeOH) λ_{max} ($\Delta\varepsilon$) 195 (−8.1), 208 (+11.2), 228 (+5.7), 258 (+3.2), 298 (+0.2) nm; ^{1}H and ^{13}C NMR, see Table 1; Table 2; HRESIMS *m/z* 418.1737 [M + Na]$^+$ (calcd for C$_{22}$H$_{25}$N$_3$O$_4$Na, 418.1737 [M + Na]$^+$).

3.5. Biological Assay

Cytotoxic Assay. The cytotoxic activity of compounds **1**–**8** were evaluated in vitro by the MTT method [30], with the concentration of 10 μM. Three human tumor cell lines were used, including human promyelocytic leukemia cell line (HL-60), human colorectal cancer cell line (HCT-116), and human breast cancer cell line (MCF-7), with cisplatinum (DDP) as a positive control.

Antibacterial Assay. The antibacterial activity of the isolated compounds was determined using the conventional broth dilution assay [31] with ciprofloxacin (CIP) as a positive control. Sixteen pathogenic bacterial strains were used, including *Bacillus megaterium*, *Bacillus subtilis*, *Escherichia coli*, *Bacillus anthraci*, *Bacillus cereus*, *Bacterium paratyphosum B*, *Enterobacter aerogenes*, *Micrococcus lysodeikticus*, *Micrococcus luteus*, *Proteusbacillm vulgaris*, *Shigella dysenteriae*, *Psmdomonas aeruginosa*, *Staphylococcus aureus*, *Salmonella typhi*, *Vibrio anguillarum*, and *Vibrio parahemolyticus*. The concentration of compounds **1**–**8** was 25 μM.

3.6. ECD Spectrum Measurement and Calculation

The tested compounds **1**–**5** were dissolved in chromatographic methanol at the concentration of 0.5 mg/mL and transferred 300 μL to quartz cuvette (1 cm for width) for tested. First tested the ECD spectrum of methanol as the blank control, then tested for compounds **1**–**5**. The wavelength was set at 190–400 nm with the band width of 1 nm, and Nitrogen should be used throughout the experiment at the flow rate of 3 mL/min. The ECD spectrum was calculated by the TDDFT methodology with a larger basis set at the B3LYP/6-311++G(2d,p) level and simulated using SpecDis 1.62 [32] according to Boltzmann distributions.

4. Conclusions

In summary, three new indole-diketopiperazine derivatives (**1–3**) showed selective cytotoxicities have been obtained from the marine-derived fungus *P. brasilianum* HBU-136 with the aid of genomic analysis. ECD quantum chemistry calculations were carried out to assign the absolute configurations of the new compounds **1–3**. The present research indicated that it is a powerful approach to discover new natural products from marine-derived fungi by the combination of chemical and bioinformatics analyses.

Supplementary Materials: The following are available online at http://www.mdpi.com/1660-3397/17/9/514/s1, NMR and MS data of **1–8**; Supplementary tables and figures for BGC analyses; tables for biological assay.

Author Contributions: Y.-H.Z. contribute to fermentation, extraction, isolation, and manuscript preparation; C.G. and X.-W.Z. contribute to genome sequencing; H.-J.Z. contribute to quantum chemistry calculation; C.-L.S. contribute to bioactivities test; F.C. and C.-Y.W. were the project leaders organizing and guiding the experiments and manuscript writing.

Funding: This work was supported by the National Natural Science Foundation of China (Nos. 41606174; 81673350; 41830535), the Marine S&T Fund of Shandong Province for Pilot National Laboratory for Marine Science and Technology (Qingdao, China) (No. 2018SDKJ0406-5), the Fundamental Research Funds for the Central Universities of China (No. 201762017), Shandong Provincial Natural Science Foundation (No. ZR2017BD009), and the High Performance Computer Center of Hebei University, and the Taishan Scholars Program, China.

Conflicts of Interest: The authors declare no conflict of interest.

References

1. Carroll, A.R.; Copp, B.R.; Davis, R.A.; Keyzers, R.A.; Prinsep, M.R. Marine natural products. *Nat. Prod. Rep.* **2019**, *36*, 122–173. [CrossRef] [PubMed]
2. Piel, J.; Hoang, K.; Moore, B.S. Natural metabolic diversity encoded by the enterocin biosynthesis gene cluster. *J. Am. Chem. Soc.* **2000**, *122*, 5415–5416. [CrossRef]
3. Xu, W.; Dhingra, S.; Chooi, Y.H.; Calvo, A.M.; Lin, H.C.; Tang, Y. The fumagillin biosynthetic gene cluster in *Aspergillus fumigatus* encodes a cryptic terpene cyclase involved in the formation of β-trans-bergamotene. *J. Am. Chem. Soc.* **2013**, *135*, 4616–4619.
4. Wang, J.F.; He, W.J.; Huang, X.L.; Tian, X.P.; Liao, S.R.; Yang, B.; Wang, F.Z.; Zhou, X.J.; Liu, Y.H. Antifungal new oxepine-containing alkaloids and xanthones from the deep-sea-derived fungus *Aspergillus versicolor* SCSIO 05879. *J. Agric. Food Chem.* **2016**, *64*, 2910–2916. [CrossRef] [PubMed]
5. Song, F.H.; Liu, X.R.; Guo, H.; Ren, B.; Chen, C.X.; Piggott, A.M.; Yu, K.; Gao, H.; Wang, Q.; Liu, M.; et al. Brevianamides with antitubercular potential from a marine-derived isolate of *Aspergillus versicolor*. *Org. Lett.* **2012**, *14*, 4770–4773. [CrossRef] [PubMed]
6. Wang, M.Z.; Sun, M.W.; Hao, H.L.; Lu, C.H. Avertoxins A–D, Prenyl asteltoxin derivatives from *Aspergillus versicolor* Y10, an endophytic fungus of *Huperzia serrate*. *J. Nat. Prod.* **2015**, *78*, 3067–3070. [CrossRef] [PubMed]
7. Zhou, M.; Miao, M.M.; Du, G.; Li, X.N.; Shang, S.Z.; Zhao, W.; Liu, Z.H.; Yang, G.Y.; Che, C.T.; Hu, Q.F.; et al. Aspergillines A–E, highly oxygenated hexacyclic indole–tetrahydrofuran–tetramic acid derivatives from *Aspergillus versicolor*. *Org. Lett.* **2014**, *16*, 5016–5019. [CrossRef]
8. Son, S.; Hong, Y.S.; Jang, M.; Heo, K.T.; Lee, B.; Jang, J.P.; Kim, J.W.; Ryoo, I.J.; Kim, W.G.; Ko, S.K.; et al. Genomics-driven discovery of chlorinated cyclic hexapeptides ulleungmycins A and B from a Streptomyces species. *J. Nat. Prod.* **2017**, *80*, 3025–3031. [CrossRef]
9. Hu, Y.Y.; Wang, M.; Wu, C.Y.; Tan, Y.; Li, J.; Hao, X.M.; Duan, Y.B.; Guan, Y.; Shang, X.Y.; Wang, Y.G.; et al. Identification and proposed relative and absolute configurations of niphimycins C–E from the marine-derived *Streptomyces* sp. IMB7-145 by genomic analysis. *J. Nat. Prod.* **2018**, *81*, 178–187. [CrossRef]
10. Xu, X.K.; Zhou, H.B.; Liu, Y.; Liu, X.T.; Fu, J.; Li, A.Y.; Li, Y.Z.; Shen, Y.M.; Bian, X.Y.; Zhang, Y.M. Heterologous Expression Guides Identification of the Biosynthetic Gene Cluster of Chuangxinmycin, an Indole Alkaloid Antibiotic. *J. Nat. Prod.* **2018**, *81*, 1060–1064. [CrossRef]
11. Zhu, A.; Zhang, X.W.; Zhang, M.; Li, W.; Ma, Z.Y.; Zhu, H.J.; Cao, F. Aspergixanthones I–K, new anti-Vibrio prenylxanthones from the marine-derived fungus *Aspergillus* sp. ZA-01. *Mar. Drugs* **2018**, *16*, 312–319. [CrossRef] [PubMed]

12. Yang, J.K.; Zhang, B.; Gao, T.; Yang, M.Y.; Zhao, G.Z.; Zhu, H.J.; Cao, F. A pair of enantiomeric 5-oxabicyclic [4.3.0] lactam derivatives and one new polyketide from the marine-derived fungus *Penicillium griseofulvum*. *Nat. Prod. Res.* **2018**, *32*, 2366–2369. [CrossRef] [PubMed]

13. Cao, F.; Meng, Z.H.; Mu, X.; Yue, Y.F.; Zhu, H.J. Absolute configuration of bioactive azaphilones from the marine-derived fungus *Pleosporales* sp. CF09-1. *J. Nat. Prod.* **2019**, *82*, 386–392. [CrossRef] [PubMed]

14. Zhu, A.; Yang, M.Y.; Zhang, Y.H.; Shao, C.L.; Wang, C.Y.; Hu, L.D.; Cao, F.; Zhu, H.J. Absolute configurations of 14, 15-hydroxylated prenylxanthones from a marine-derived *Aspergillus* sp. fungus by chiroptical methods. *Sci. Rep.* **2018**, *8*, 10621–10630. [CrossRef] [PubMed]

15. Wang, F.Z.; Fang, Y.C.; Zhu, T.J.; Zhang, M.; Lin, A.Q.; Gu, Q.Q.; Zhu, W.M. Seven new prenylated indole diketopiperazine alkaloids from holothurian-derived fungus *Aspergillus fumigatus*. *Tetrahedron* **2008**, *64*, 7986–7991. [CrossRef]

16. Cui, C.B.; Kakeya, H.; Osada, H. Novel mammalian cell cycle inhibitors, spirotryprostatins A and B, produced by *Aspergillus fumigatus*, which inhibit mammalian cell cycle at G2/M phase. *Tetrahedron* **1996**, *52*, 12651–12666. [CrossRef]

17. Kato, N.; Suzuki, H.; Takagi, H.; Asami, Y.; Kakeya, H.; Uramoto, M.; Usui, T.; Takahashi, S.; Sugimoto, Y.; Osada, H. Identification of cytochrome P450s required for fumitremorgin biosynthesis in *Aspergillus fumigatus*. *ChemBioChem.* **2010**, *10*, 920–928. [CrossRef] [PubMed]

18. Maiya, S.; Grundmann, A.; Li, S.M.; Turner, G. The fumitremorgin gene cluster of *Aspergillus fumigatus*: Identification of a gene encoding brevianamide F synthetase. *ChemBioChem.* **2010**, *7*, 1062–1069. [CrossRef]

19. Grundmann, A.; Kuznetsova, T.; Afiyatullov, S.S.; Li, S.M. FtmPT2, an *N*-Prenyltransferase from *Aspergillus fumigatus*, catalyses the last step in the biosynthesis of fumitremorgin B. *ChemBioChem* **2008**, *9*, 2059–2063. [CrossRef]

20. Tsunematsu, Y.; Ishiuchi, K.; Hotta, K.; Watanabe, K. Yeast-based genome mining, production and mechanistic studies of the biosynthesis of fungal polyketide and peptide natural products. *Nat. Prod. Rep.* **2013**, *30*, 1139–1149. [CrossRef]

21. Gu, B.; He, S.; Yan, X.; Zhang, L. Tentative biosynthetic pathways of some microbial diketopiperazines. *Appl. Microbiol. Biot.* **2013**, *97*, 8439–8453. [CrossRef] [PubMed]

22. Tsunematsu, Y.; Ishikawa, N.; Wakana, D.; Goda, Y.; Noguchi, H.; Moriya, H.; Hotta, K.; Watanabe, K. Distinct mechanisms for spiro-carbon formation reveal biosynthetic pathway crosstalk. *Nat. Chem. Biol.* **2013**, *9*, 818–827. [CrossRef] [PubMed]

23. Zhang, X.W.; Li, S.Y. Expansion of chemical space for natural products by uncommon P450 reactions. *Nat. Prod. Rep.* **2017**, *34*, 1061–1089. [CrossRef] [PubMed]

24. Wang, W.J.; Li, D.Y.; Li, Y.C.; Hua, H.M.; Ma, E.L.; Li, Z.L. Caryophyllene sesquiterpenes from the marine-derived fungus *Ascotricha* sp. ZJ-M-5 by the one strain–many compounds strategy. *J. Nat. Prod.* **2014**, *77*, 1367–1371. [CrossRef] [PubMed]

25. Wang, Y.; Li, Z.L.; Bai, J.; Zhang, L.M.; Wu, X.; Zhang, L.; Pei, Y.H.; Jing, Y.K.; Hua, H.M. 2,5-Diketopiperazines from the Marine-Derived Fungus *Aspergillus fumigatus* YK-7. *Chem. Biodivers.* **2012**, *9*, 385–393. [CrossRef]

26. Yu, L.Y.; Ding, W.J.; Wang, Q.Q.; Ma, Z.J.; Xu, X.W.; Zhao, X.F.; Chen, Z. Induction of cryptic bioactive 2, 5-diketopiperazines in fungus *Penicillium* sp. DT-F29 by microbial co-culture. *Tetrahedron* **2017**, *73*, 907–914. [CrossRef]

27. Xue, H.; Lu, C.H.; Liang, L.Y.; Shen, Y.M. Secondary Metabolites of *Aspergillus* sp. CM9a, an Endophytic Fungus of *Cephalotaxus mannii*. *Rec. Nat. Prod.* **2012**, *6*, 28–34.

28. Wang, F.Q.; Tong, Q.Y.; Ma, H.R.; Xu, H.F.; Hu, S.; Ma, W.; Xue, Y.B.; Liu, J.J.; Wang, J.P.; Song, H.P.; et al. Indole diketopiperazines from endophytic *Chaetomium* sp. 88194 induce breast cancer cell apoptotic death. *Sci. Rep.* **2015**, *5*, 9294. [CrossRef]

29. Meng, L.H.; Du, F.Y.; Li, X.M.; Pedpradab, P.; Xu, G.M.; Wang, B.G. Rubrumazines A–C, indolediketopiperazines of the isoechinulin class from *Eurotium rubrum* MA-150, a fungus obtained from marine mangrove-derived rhizospheric soil. *J. Nat. Prod.* **2015**, *78*, 909–913. [CrossRef]

30. Scudiere, D.A.; Shoemaker, R.H.; Paull, K.D.; Monks, A.; Tierney, S.; Nofziger, T.H.; Currens, M.J.; Seniff, D.; Boyd, M.R. Evaluation of a soluble tetrazolium/formazan assay for cell growth and drug sensitivity in culture using human and other tumor cell lines. *Cancer Res.* **1988**, *48*, 4827–4833.

31. Appendino, G.; Gibbons, S.; Giana, A.; Pagani, A.; Grassi, G.; Stavri, M.; Smith, E.; Rahman, M.M. Antibacterial cannabinoids from Cannabis sativa: A structure−Activity study. *J. Nat. Prod.* **2008**, *71*, 1427–1430. [CrossRef] [PubMed]

32. Bruhn, T.; Schaumlöffel, A.; Hemberger, Y.; Bringmann, G. SpecDis: Quantifying the comparison of calculated and experimental electronic circular dichroism spectra. *Chirality* **2013**, *25*, 243–249. [CrossRef] [PubMed]

Article

Two New Spiro-Heterocyclic γ-Lactams from A Marine-Derived *Aspergillus fumigatus* Strain CUGBMF170049

Xiuli Xu [1,*], Jiahui Han [1,2], Yanan Wang [1], Rui Lin [1], Haijin Yang [1], Jiangpeng Li [1], Shangzhu Wei [1], Steven W. Polyak [3] and Fuhang Song [2,*]

[1]   School of Ocean Sciences, China University of Geosciences, Beijing 100083, China;
      15632779760@163.com (J.H.); nancywang0410@163.com (Y.W.); linrui520@126.com (R.L.);
      yanghaijin52@163.com (H.Y.); cuculjp2016@163.com (J.L.); wert0715@163.com (S.W.)
[2]   CAS Key Laboratory of Pathogenic Microbiology and Immunology, Institute of Microbiology,
      Chinese Academy of Sciences, Beijing 100101, China
[3]   School of Pharmacy and Medical Sciences, University of South Australia, Adelaide 5000, Australia;
      steven.polyak@unisa.edu.au
*   Correspondence: xuxl@cugb.edu.cn (X.X.); songfuhang@im.ac.cn (F.S.); Tel.: +86-10-8231-9124 (X.X.);
      +86-10-6480-6058 (F.S.)

Received: 18 April 2019; Accepted: 8 May 2019; Published: 14 May 2019

Abstract: Two new spiro-heterocyclic γ-lactam derivatives, cephalimysins M (**1**) and N (**2**), were isolated from the fermentation cultures of the marine-derived fungus *Aspergillus fumigatus* CUGBMF17018. Two known analogues, pseurotin A (**3**) and FD-838 (**4**), as well as four previously reported helvolic acid derivatives, 16-*O*-propionyl-16-*O*-deacetylhelvolic acid (**5**), 6-*O*-propionyl-6-*O*-deacetylhelvolic acid (**6**), helvolic acid (**7**), and 1,2-dihydrohelvolic acid (**8**) were also identified. One-dimensional (1D), two-dimensional (2D) NMR, HRMS, and circular dichroism spectral analysis characterized the structures of the isolated compounds.

Keywords: marine-derived *Aspergillus fumigatus*; spiro-heterocyclic γ-lactam; cephalimysins

1. Introduction

Marine-derived fungi are important resources of structurally and biologically diverse secondary metabolites in drug discovery [1–6]. A series of novel marine natural compounds have been isolated from marine-derived fungi of *Aspergillus fumigatus* strains, such as *E*-β-*trans*-5,8,11-trihydroxybergamot-9-ene and β-*trans*-2β,5,15-trihydroxybergamot-10-ene [7], diketopiperazines [8,9], indole alkaloids [10], fumigaclavine C [11], fumiquinazoline K [12], and gliotoxin analogues [13]. During our ongoing efforts to search for new bioactive metabolites from marine-derived fungi, an *Aspergillus fumigatus* strain CUGBMF170049 was isolated from a sediment sample that was collected from the Bohai Sea, China. Chemical investigations on an EtOAc-MeOH extracted fraction of its solid fermentation cultures resulted in the characterization of two new spiro-heterocyclic γ-lactam derivatives, cephalimysins M (**1**) and N (**2**), along with two known analogues, pseurotin A (**3**) [14], and FD-838 (**4**) [15], as well as four previously reported helvolic acid derivatives, 16-*O*-propionyl-16-*O*-deacetylhelvolic acid (**5**), 6-*O*-propionyl-6-*O*-deacetylhelvolic acid (**6**) [16], helvolic acid (**7**), and 1,2-dihydrohelvolic acid (**8**) [17] (Figure 1). Herein, we report the isolation and structural determination of the new compounds **1** and **2**. The antibacterial activities of the compounds were also investigated against a panel of both Gram-positive and Gram-negative bacteria, *Mycobacterium* bovis bacillus Calmette Guérin (BCG) and *Candida albicans*. Compounds **5–7** showed significant antibacterial activities against both *Staphylococcus aureus* and methicillin resistant *S. aureus*.

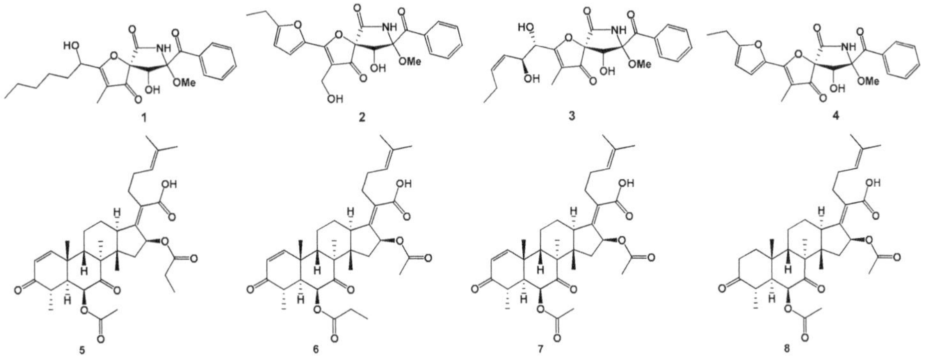

Figure 1. Chemical structures of **1–8**.

2. Results

2.1. Structure Elucidation

Compound **1** was obtained as pale yellow amorphous powder. Its molecular formula was determined as $C_{22}H_{27}NO_7$ by HRESIMS *m/z* 440.1684 [M + Na]$^+$ (calcd. for $C_{22}H_{27}NO_7Na$ 440.1680, Δmmu + 0.4) (Figure S1 in Supplementary Materials), which accounted for ten degrees of unsaturation. The ^1H, ^{13}C, and HSQC NMR spectra (Figures S2–S4 in Supplementary Materials) of compound **1** showed signals of two ketone carbonyls at δ_C 196.7 (C-4) and 196.4 (C-17), one amide carbonyl at δ_C 166.4 (C-6), a mono-substituted benzene ring (δ_C 133.4, δ_C 130.3/δ_H 8.25, δ_C 128.4/δ_H 7.53, δ_C 133.9/δ_H 7.64, δ_C 128.4/δ_H 7.53, 130.3/δ_H 8.25), two sp^2 quaternary carbons at δ_C 188.5 (C-2) and 109.6 (C-3), four sp^3 methylenes (δ_C 34.4/δ_H 1.64, δ_C 31.1/δ_H 1.24, δ_C 24.1/δ_H 1.37, and δ_C 22.0/δ_H 1.24), two sp^3 oxygenated methines (δ_C 74.9/δ_H 4.38 and δ_C 67.0/δ_H 4.50), two sp^3 oxygenated quaternary carbons at δ_C 91.2 (C-5) and δ_C 92.5 (C-8), and three methyls (δ_C 13.9/δ_H 0.84, δ_C 5.4/δ_H 1.64, and δ_C 51.7/δ_H 3.24) (Table 1). A comparison of ^1H and ^{13}C NMR data for compound **1** (Table 1) with those of previously reported pseurotin A (**3**) [14] revealed many similarities in their chemical structures, except for the oxygenated unsaturated side chain of **3** had been substituted with an oxygenated saturated fatty chain in **1**. Compound **1** had the same skeleton as that of pseurotin A, while the side chain of **1** is an oxygenated saturated fatty chain. The COSY correlations (Figure 2, and Figure S5 in Supplementary Materials) from H-19 (δ_H 8.25) to H-23 (δ_H 8.25), through H-20 (δ_H 7.53), H-21 (δ_H 7.64) and H-22 (δ_H 7.53) identified the mono-substituted benzene ring. Likewise, the fatty acid side chain was confirmed by COSY correlations from H-10 (δ_H 4.50) to H$_3$-15 (δ_H 0.84), through H-11 (δ_H 1.64), H-12 (δ_H 1.37), H-13 (δ_H 1.24), and H-14 (δ_H 1.24). The lactam ring was evidenced by the HMBC correlations (Figure 2, and Figure S6 in Supplementary Materials) from H-9-OH (δ_H 6.34) to C-5 (δ_C 91.2), C-8 (δ_C 92.5), and C-9 (δ_C 74.9), as well as from H-7-NH (δ_H 9.90) to C-5 (δ_C 91.2), C-6 (δ_C 166.4), C-8 (δ_C 92.5), and C-9 (δ_C 74.9). The phenylmethanone moiety was confirmed by the HMBC crossing peaks from H-19 (δ_H 8.25) and H-23 (δ_H 8.25) to C-17 (δ_C 196.4), and the connection from C-8 (δ_C 92.5) to C-17 (δ_C 196.4) was revealed by the HMBC correlation from H-9 (δ_H 4.38) to C-17 (δ_C 196.4). The HMBC correlation from H$_3$-24 (δ_H 3.24) to C-8 (δ_C 92.5) confirmed the methoxy at C-8. The connection from C-10 to C-4 through C-2 and C-3 was confirmed by the HMBC correlations from H-10-OH (δ_H 5.62) to C-10 (δ_C 67.0) and C-2 (δ_C 188.5), from H-10 (δ_H 4.50) to C-2 (δ_C 188.5) and C-3 (δ_C 109.6), as well as from H$_3$-16 (δ_H 1.64) to C-2 (δ_C 188.5), C-3 (δ_C 109.6), and C-4 (δ_C 196.7). In addition, the spirobicyclic moiety was suggested by the HMBC correlation from H-9 (δ_H 4.38) to C-4 (δ_C 196.7), the chemical shift of C-5 (δ_C 91.2) and the molecular formula. The *cis* configurations of 8-OCH$_3$ and 9-OH were supported by the chemical shift of H-9 and the coupling constant (δ_H 4.38, *J* = 9.0 Hz) between H-9 and 9-OH [15,18]. The circular dichroism (CD) spectrum (Figure S7 in Supplementary Materials) of **1** showed negative Cotton effects at around 230, 280, and 345 nm, and positive Cotton effects at around

250 and 310 nm, which were consistent with the reported CD data for pseurotin A [18]. Thus, the structure of compound **1** was established, as shown in Figure 1, and was named cephalimysin M, its absolute configurations for C-5, C-8, and C-9 being assigned the same as those of pseurotin A. The absolute configuration of C-10 was not defined.

Table 1. NMR data for **1** and **2** (DMSO-d_6).

Position	1		2	
	δ_C, Type	δ_H, mult (*J* in Hz)	δ_C, Type	δ_H, mult (*J* in Hz)
2	188.5, C		172.2, C	
3	109.6, C		111.6, C	
4	196.7, C		193.4, C	
5	91.2, C		92.1, C	
6	166.4, C		166.4, C	
8	92.5, C		92.6, C	
9	74.9, CH	4.38, d (9.0)	75.0, CH	4.50, s
10	67.0, CH	4.50, td (7.2, 5.4)	141.8, C	
11	34.4, CH$_2$	1.64, m (overlap)	120.0, CH	7.40, d (3.6)
12	24.1, CH$_2$	1.37, m	108.6, CH	6.52, d (3.6)
13	31.1, CH$_2$	1.24, m	163.4, C	
14	22.0, CH$_2$	1.24, m	21.0, CH$_2$	2.76, q (7.8)
15	13.9, CH$_3$	0.84, t (7.2)	11.8, CH$_3$	1.22, t (7.8)
16	5.4, CH$_3$	1.64, s	50.2, CH$_2$	4.25, s
17	196.4, C		196.4, C	
18	133.4, C		133.4, C	
19	130.3, CH	8.25, dd (8.4, 1.2)	130.4, CH	8.27, d (7.2)
20	128.4, CH	7.53, ddd (8.4, 8.4, 1.2)	128.4, CH	7.53, dd (7.2, 7.2)
21	133.9, CH	7.64, dddd (8.4, 8.4, 1.2, 1.2)	133.9, CH	7.67, dd (7.2, 7.2)
22	128.4, CH	7.53, ddd (8.4, 8.4, 1.2)	128.4, CH	7.53, dd (7.2, 7.2)
23	130.3, CH	8.25, dd (8.4, 1.2)	130.4, CH	8.27, d (7.2)
24	51.7, CH$_3$	3.24, s	51.7, CH$_3$	3.26, s
7-NH		9.90, s		9.98, s
9-OH		6.34, d (9.0)		
10-OH		5.62, d (5.4)		

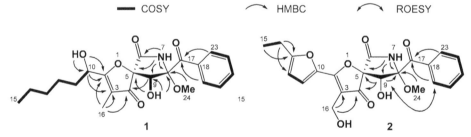

Figure 2. Key two-dimensional (2D) NMR correlations for **1** and **2**.

Compound **2** was obtained as pale yellow amorphous powder. Its molecular formula was determined as C$_{22}$H$_{21}$NO$_8$ by HRESIMS *m/z* 450.1159 [M + Na]$^+$ (calcd. for C$_{22}$H$_{21}$NO$_8$Na 450.1159, Δmmu 0) (Figure S8 in Supplementary Materials), which accounted for thirteen degrees of unsaturation. The ^{1}H, ^{13}C, and HSQC NMR spectra (Figures S9–S11 in Supplementary Materials) of compound **2** showed signals of two ketone carbonyls at δ_C 193.4 (C-4) and 196.4 (C-17), one amide carbonyl at δ_C 166.4 (C-6), a mono-substituted benzene ring (δ_C 133.4, δ_C 130.4/δ_H 8.27, δ_C 128.4/δ_H 7.53, δ_C 133.9/δ_H 7.67, δ_C 128.4/δ_H 7.53, δ_C 130.4/δ_H 8.27), two sp^2 methines (δ_C 120.1/δ_H 7.40, δ_C 108.6/δ_H 6.52), four sp^2 quaternary carbons at δ_C 172.2 (C-2), 111.6 (C-3), 141.8 (C-10), and 163.4 (C-13), one sp^3 methylene (δ_C 21.0/δ_H 2.76), one sp^3 oxygenated methylene (δ_C 50.2/δ_H 4.25), one sp^3 oxygenated methine

(δ_C 75.0/δ_H 4.50), two sp^3 oxygenated quaternary carbons (δ_C 92.1 and δ_C 92.6), and two methyls (δ_C 11.8/δ_H 1.22 and δ_C 51.7/δ_H 3.26) (Table 1). A comparison of ^{1}H and ^{13}C NMR data for compound **2** (Table 1) with that of the previously reported FD-838 (**4**) [15] revealed many similarities. Compound **2** had the same skeleton as that of FD-838, while methyl at C-3 of compound **4** was replaced by an oxygenated methylene in compound **2**. The mono-substituted benzene ring was identified by the COSY correlations (Figure 2, and Figure S12 in Supplementary Materials) from H-19 (δ_H 8.27) to H-23 (δ_H 8.27), through H-20 (δ_H 7.53), H-21 (δ_H 7.67), and H-22 (δ_H 7.53), and the furan side chain was characterized by COSY correlations between H-11 (δ_H 7.40) and H-12 (δ_H 6.52), and between H$_2$-14 (δ_H 2.76) and H$_3$-15 (δ_H 1.22), along with the HMBC correlations (Figure 2, and Figure S13 in Supplementary Materials) from H$_3$-15 to C-13, and from H$_2$-14 to C-12 and C-13. The lactam ring was suggested by the HMBC correlations from H-9 (δ_H 4.50) to C-5 (δ_C 92.1) and C-8 (δ_C 92.6), as well as from H-7-NH (δ_H 9.98) to C-5 (δ_C 92.1), C-8 (δ_C 92.6) and C-9 (δ_C 75.0). HMBC crossing peaks from H-19 (δ_H 8.27) and H-23 (δ_H 8.27) to C-17 (δ_C 196.4) confirmed the phenylmethanone moiety, and the connection from C-8 (δ_C 92.6) to C-17 (δ_C 196.4) was revealed by the HMBC correlation from H-9 (δ_H 4.50) to C-17 (δ_C 196.4). The methoxy at C-8 was also confirmed by the HMBC correlation from H$_3$-24 (δ_H 3.26) to C-8 (δ_C 92.6). The spirobicyclic moiety was indicated by the HMBC correlation from H-9 (δ_H 4.50) to C-4 (δ_C 193.4) and the chemical shift of C-5 (δ_C 92.1). The ROESY correlations (Figure 2, and Figure S14 in Supplementary Materials) from H-9 (δ_H 4.50) to H-19 (δ_H 8.27) and H-23 (δ_H 8.27) indicated the relative configurations of C-8 and C-9. The chemical shifts of C-5, C-8, and C-9 for **2** were much closer to those of **1** and FD-838 [15], which defined the relative configurations of C-5, C-8, and C-9. The circular dichroism (CD) spectrum (Figure S15 in Supplementary Materials) of **2** showed a negative Cotton effect at around 318 nm and a positive Cotton effect at around 355 nm, which were consistent with the reported CD data for FD-838 [15]. Therefore, the structure of compound **2** was established, as shown in Figure 1, where absolute configurations were assigned the same as those of FD-838 and named cephalimysin N.

In addition to compounds **1** and **2**, known compounds were also identified in the fermentation products, such as pseurotin A (**3**) [14], FD-838 (**4**) [15], as well as four known helvolic acid derivatives, 16-*O*-propionyl-16-*O*-deacetylhelvolic acid (**5**), 6-*O*-propionyl-6-*O*-deacetylhelvolic acid (**6**) [16], helvolic acid (**7**), and 1,2-dihydrohelvolic acid (**8**) [17].

2.2. Biological Activity

Compounds **1**–**8** were tested against Gram positive bacteria *S. aureus* (ATCC 6538) and methicillin resistant *S. aureus* (MRSA) (ATCC 29213), Gram negative bacteria *Escherichia coli* (ATCC 11775) and *Pseudomonas aeruginosa* (ATCC 15692), BCG, and *C. albicans*. Compounds **5**–**7** showed antibacterial activities against both *S. aureus* and MRSA. Comparing the antibacterial activities of compounds **5**–**7** with the inactive analogue **8** indicated that the α,β-unsaturated ketone appears to be a key functional group for antibacterial activity (Table 2). None of the isolated compounds exhibited antimicrobial activities against *E. coli*, *P. aeruginosa*, *C. albicans* (MIC > 100 μg/mL), nor BCG (MIC > 10 μg/mL).

Table 2. Antimicrobial activities of **1**–**8** (μg/mL).

Compounds	*S. aureus* [a]	MRSA [a]
1	>100	>100
2	>100	>100
3	>100	>100
4	>100	>100
5	12.5	25
6	6.25	12.5
7	0.78	0.78
8	>100	>100

[a] Vancomycin was used as positive control with MIC value of 0.78 μg/mL.

3. Materials and Methods

3.1. General Experimental Procedures

The optical rotations ($[\alpha]_D$) were measured on Anton Paar MCP 200 Modular Circular Polarimeter (Austria) in a 100×2 mm cell at 22 °C. CD spectra were recorded on an Applied Photophysics Chirascan spectropolarimeter (UK). NMR spectra were obtained on a Bruker Avance DRX600 spectrometer with residual solvent peaks serving as references (DMSO-d_6: δ_H 2.50, δ_C 39.52). High-resolution ESIMS measurements were obtained on a Bruker micrOTOF mass spectrometer by direct infusion in MeCN at 3 mL/min using sodium formate clusters as an internal calibrate. HPLC was performed using an Agilent 1200 Series separation module that was equipped with Agilent 1200 Series diode array and Agilent 1260 Series fraction collector, and Agilent SB-C18 column (250×9.4 mm, 5 µm).

3.2. Fungal Material

The *Aspergillus fumigatus* strain CUGBMF170049 was isolated from a sediment sample that was collected from the Bohai Sea, China and grown on a potato dextrose agar plate at 28 °C. This strain was identified as *Aspergillus fumigatus* based on DNA sequence analysis of its internal transcribed spacer (ITS) region (Figure S16) (GenBank accession number MK453215) using a conventional primer pair of ITS5 (5′-GGAAGTAAAAGTCGTAACAAGG-3′) and ITS4 (5′-TCCTCCGCTTATTGATATGC-3′).

3.3. Fermenttion and Extraction

A small spoonful of *Aspergillus fumigatus* (CUGBMF170049) spores growing on a potato dextrose agar slant was inoculated into four 250 mL conical flasks, each containing 40 mL of liquid medium consisting of potato infusion (20%), glucose (2.0%), artificial sea salt (3.5%), and distilled water. The flasks were incubated at 28 °C for 3 d on a rotary shaker at 160 rpm. An aliquot (5 mL) of the resultant seed culture was inoculated into teen 1 L conical flasks, with each containing solid medium consisting of rice (120 g) and artificial seawater (3.5%; 80 mL), and the flasks were incubated stationary for 30 days at 28 °C. The cultures were extracted three times by EtOAc:MeOH (80:20), and the combined extracts were reduced to dryness in vacuo and the residue was partitioned between EtOAc and H_2O. Subsequently, the EtOAc layer was dried in vacuo to yield a dark residue (11.3 g).

3.4. Isolation and Purification

The EtOAc fraction was fractionated by a reduced pressure silica gel chromatography (50×80 mm column, TLC H silica) using a stepwise gradient of 50–100% hexane/CH_2Cl_2 and then 0–100% MeOH/CH_2Cl_2 to afford 15 fractions. Fraction C was fractionated on a Sephadex LH-20 column (600×30 mm) while using an isocratic elution of hexane:CH_2Cl_2:MeOH (5:5:1) to give five subfractions (F1–F5). Subfraction F3 (102.3 mg after drying in vacuo) was further fractionated by HPLC (Agilent Zorbax SB-C18 250×9.4 mm, 5 µm column, 2.0 mL/min, isocratic 65% MeOH/H_2O) to yield FD-838 (**4**; t_R 10.4 min, 3.3 mg). Fraction J was fractionated on a Sephadex LH-20 column (600×30 mm) using an isocratic elution of CH_2Cl_2:MeOH (2:1), to give four subfractions (F1–F4). Subfraction F1 was further fractionated by HPLC (Agilent Zorbax SB-C18 250×9.4 mm, 5 µm column, 2.0 mL/min, isocratic 65% MeOH/H_2O) to yield helvolic acid (**7**, t_R 10.8 min, 1.3 mg), 16-*O*-propionyl-16-*O*-deacetylhelvolic acid (**5**, t_R 11.9 min, 1.2 mg), and 6-*O*-propionyl-6-*O*-deacetylhelvolic acid (**6**, t_R 12.4 min, 1.4 mg). Fraction K was fractionated on a Sephadex LH-20 column (600×30 mm) using an isocratic elution of CH_2Cl_2:MeOH (2:1) to give five subfractions (F1–F5). Subfraction F3 was further fractionated by HPLC (Agilent Zorbax SB-C18 250×9.4 mm, 5 µm column, 2.0 mL/min, isocratic 65% MeOH/H_2O) to yield 1,2-dihydrohelvolic acid (**8**, t_R 13.9 min, 1.6 mg). Fraction L was fractionated on a Sephadex LH-20 column (600×30 mm) using an isocratic elution of CH_2Cl_2:MeOH (2:1) to give five subfractions (F1–F5). Subfraction F4 was further fractionated by an ODS column, which was eluted by a stepwise gradient (0–100% MeOH/H_2O) to afford five subfractions (F1–F5). Subfraction F4 was further fractionated by HPLC (Agilent Zorbax SB-C18 250×9.4 mm, 5 µm column, 2.0 mL/min, isocratic 65% MeOH/H_2O) to yield pseurotin A (**3**, t_R 7.1 min, 3.2 mg), cephalimysins M (**1**, t_R 12.0 min, 1.5 mg), and N (**2**, t_R 8.9 min, 3.6 mg).

3.4.1. Cephalimysin M (**1**)

Pale yellow amorphous powder; $[\alpha]_D^{22}$ –21.3 (MeOH, 0.1); UV (MeOH) λ_{max} (logε) 196 (4.43), 254 (4.14), 277(3.96) nm; (+)-ESIMS *m/z* 418.1 [M + H]$^+$; (+)-HRESIMS *m/z* 440.1684 [M + Na]$^+$ (calcd. for $C_{22}H_{27}NO_7Na$ 440.1680); ^{1}H and ^{13}C NMR data: See Table 1.

3.4.2. Cephalimysin N (**2**)

Pale yellow amorphous powder; $[\alpha]_D^{22}$ –21.5 (MeOH, 0.1); UV (MeOH) 197 (4.43), 252 (4.12), 329(3.56) nm; (+)-ESIMS *m/z* 428.0 [M + H]$^+$; (+)-HRESIMS *m/z* 450.1159 [M + Na]$^+$ (calcd. for $C_{22}H_{21}NO_8Na$ 450.1159); ^{1}H and ^{13}C NMR data: See Table 1.

3.5. Antimicrobial Assays

The antimicrobial assays were performed according to the Antimicrobial Susceptibility Testing Standards that were outlined by the Clinical and Laboratory Standards Institute (CLSI) against *S. aureus* ATCC 6538, MRSA ATCC 29213, *E. coli* ATCC 11775, *P. aeruginosa* ATCC 15692, and *C. albicans* ATCC 10231 based on a 96 well microplate format in liquid growth. Briefly, the bacteria from glycerol stocks was inoculated on LB agar plate and cultured overnight at 37°C. The glycerol stock of *C. albicans* was prepared on Sabouraud dextrose agar at 28 °C for 24 h. Afterwards, single colonies were picked and adjusted to approximately 10^4 CFU/mL with Mueller–Hinton Broth as bacterial suspension and with RPMI 1640 media as fungal suspension. 2 μL of two-fold serial dilution of each compound (in DMSO) were added to each row on 96-well microplate, containing 78 μL of bacterial or fungal suspension in each well. (Vancomycin and Ciprofloxacin were used as positive controls; Amphotericin B was used as positive for fungi; DMSO as negative control). The 96-well plate was aerobically incubated at 37 °C for 16 h. The 96-well plate of antifungal was aerobically incubated at 35 °C for 24 h. Here, MIC is defined as the minimum concentration of compound at which no bacterial growth is observed.

3.6. Anti-Bacillus Calmette Guérin (BCG) Assay

The anti-BCG assay was carried out by using a constitutive GFP expression strain (pUV3583c-GFP), according to previous published procedure (isoniazid was used as positive control with MIC value of 0.05 μg/mL) [19]. The concentrations for the tested compounds were from 0.156 to 10 μg/mL by using two-fold diluted solutions.

4. Conclusions

As part of our ongoing research program to discover novel secondary metabolites from the marine environment, eight compounds were isolated from the rice solid medium culture of the marine derived fungus CUGBMF170049 isolated from a sediment sample that was collected from the Bohai Sea, China. Two novel compounds (**1** and **2**) were isolated and characterized along with the previously reported analogues pseurotin A (**3**) and FD-838 (**4**), as well as four known helvolic acid derivatives, namely 16-*O*-propionyl-16-*O*-deacetylhelvolic acid (**5**), 6-*O*-propionyl-6-*O*-deacetylhelvolic acid (**6**), helvolic acid (**7**), and 1,2-dihydrohelvolic acid (**8**). All of the structures were confirmed by detailed analysis of the spectroscopic data. Compounds **5**–**7** showed antibacterial activity against both *S. aureus* and MRSA. Aanalogue **8** did not exhibit antibacterial activities that indicated that the α,β-unsaturated ketone of **5**–**7** is the key functional group for antibacterial activity. Structurally, cephalimysins M (**1**) and N (**2**) belong to a family of rare natural products with diverse biological activities, which contain an unusual spiro-heterocyclic γ-lactam core. To the best of our knowledge, 28 natural products of this family have been reported, including pseurotin A [14], 8-*O*-demethylpseurotin A [20] pseurotins A1 and A2 [18,21], pseurotins B – E [22], pseurotins F1 and F2 [23], 14-norpseurotin A [24] synerazol [25], azaspirene [26], azaspirofurans A and B [27], and FD-838 and cephalimysins A–L [15,28,29]. **2** is the first cephalimysin analogue where the methyl of C-16 was oxidized to hydroxymethyl. The current research diversifies the structures of this class of natural products.

Supplementary Materials: The following are available online at http://www.mdpi.com/1660-3397/17/5/289/s1, Figures S1–S15: The HRESIMS, UV, 1 D, 2D NMR, and CD spectra of compounds **1** and **2**, Figure S16 the phylogenetic tree of strain CUGBMF170049.

Author Contributions: Data curation, X.X.; Investigation, J.H., Y.W., R.L. and H.Y.; Supervision, X.X. and F.S.; Writing—original draft, X.X. and F.S.; Writing—review & editing, X.X., R.L., J.L., S.W., S.W.P. and F.S.

Funding: This work was supported in part by the National Key Research and Development Program of China (2018YFC0311002, 2017YFD0201203, 2017YFC1601300), the National Natural Science Foundation of China (31600136). SWP was supported by the National Health and Medical Research Council of Australia (GN1147538).

Conflicts of Interest: The authors declare no conflict of interest.

References

1. Liu, S.Z.; Yan, X.; Tang, X.X.; Lin, J.G.; Qiu, Y.K. New bis-alkenoic acid derivatives from a marine-derived fungus *Fusarium solani* H915. *Mar. Drugs* **2018**, *16*, 483. [CrossRef]
2. Wang, N.; Li, C.W.; Cui, C.B.; Cai, B.; Xu, L.L.; Zhu, H.J. Four new antitumor metabolites isolated from a mutant 3-f-31 strain derived from *Penicillium purpurogenum* G59. *Eur. J. Med. Chem.* **2018**, 548–558. [CrossRef] [PubMed]
3. Rai, M.; Gade, A.; Zimowska, B.; Ingle, A.P.; Ingle, P. Marine-derived phoma-the gold mine of bioactive compounds. *Appl. Microbiol. Biotechnol.* **2018**, *102*, 9053–9066. [CrossRef]
4. Zhu, A.; Zhang, X.W.; Zhang, M.; Li, W.; Ma, Z.Y.; Zhu, H.J.; Cao, F. Aspergixanthones I-K, new anti-vibrio prenylxanthones from the marine-derived fungus *Aspergillus* sp. ZA-01. *Mar. Drugs* **2018**, *16*, 312. [CrossRef]
5. Su, D.; Ding, L.; He, S. Marine-derived *Trichoderma* species as a promising source of bioactive secondary metabolites. *Mini. Rev. Med. Chem.* **2018**, *18*, 1702–1713. [CrossRef]
6. Zhang, P.; Jia, C.; Lang, J.; Li, J.; Luo, G.; Chen, S.; Yan, S.; Liu, L. Mono- and dimeric naphthalenones from the marine-derived fungus *Leptosphaerulina chartarum* 3608. *Mar. Drugs* **2018**, *16*, 173. [CrossRef] [PubMed]
7. Wang, Y.; Li, D.H.; Li, Z.L.; Sun, Y.J.; Hua, H.M.; Liu, T.; Bai, J. Terpenoids from the marine-derived fungus *Aspergillus fumigatus* YK-7. *Molecules* **2016**, *21*, 31. [CrossRef]
8. Wang, Y.; Li, Z.L.; Bai, J.; Zhang, L.M.; Wu, X.; Zhang, L.; Pei, Y.H.; Jing, Y.K.; Hua, H.M. 2,5-diketopiperazines from the marine-derived fungus *Aspergillus fumigatus* YK-7. *Chem. Biodivers.* **2012**, *9*, 385–393. [CrossRef] [PubMed]
9. Zhao, W.Y.; Zhu, T.J.; Fan, G.T.; Liu, H.B.; Fang, Y.C.; Gu, Q.Q.; Zhu, W.M. Three new dioxopiperazine metabolites from a marine-derived fungus *Aspergillus fumigatus* Fres. *Nat. Prod. Res.* **2010**, *24*, 953–957. [CrossRef]
10. Zhang, D.; Satake, M.; Fukuzawa, S.; Sugahara, K.; Niitsu, A.; Shirai, T.; Tachibana, K. Two new indole alkaloids, 2-(3,3-dimethylprop-1-ene)-costaclavine and 2-(3,3-dimethylprop-1-ene)-epicostaclavine, from the marine-derived fungus *Aspergillus fumigatus*. *J. Nat. Med.* **2012**, *66*, 222–226. [CrossRef]
11. Li, Y.X.; Himaya, S.W.; Dewapriya, P.; Zhang, C.; Kim, S.K. Fumigaclavine C from a marine-derived fungus *Aspergillus fumigatus* induces apoptosis in MCF-7 breast cancer cells. *Mar. Drugs* **2013**, *11*, 5063–5086. [CrossRef]
12. Afiyatullov, S.S.; Zhuravleva, O.I.; Antonov, A.S.; Kalinovsky, A.I.; Pivkin, M.V.; Menchinskaya, E.S.; Aminin, D.L. New metabolites from the marine-derived fungus *Aspergillus fumigatus*. *Nat. Prod. Commun.* **2012**, *7*, 497–500. [CrossRef]
13. Zhao, W.Y.; Zhu, T.J.; Han, X.X.; Fan, G.T.; Liu, H.B.; Zhu, W.M.; Gu, Q.Q. A new gliotoxin analogue from a marine-derived fungus *Aspergillus fumigatus* Fres. *Nat. Prod. Res.* **2009**, *23*, 203–207. [CrossRef]
14. Bloch, P.; Tamm, C. Isolation and structure of Pseurotin A, a microbial metabolite of *Pseudeurotium* ovalis STOLK with an unusual heterospirocyclic system. *Helv. Chim. Acta* **1981**, *64*, 304–315. [CrossRef]
15. Yamada, T.; Kitada, H.; Kajimoto, T.; Numata, A.; Tanaka, R. The relationship between the CD Cotton effect and the absolute configuration of FD-838 and its seven stereoisomers. *J. Org. Chem.* **2010**, *75*, 4146–4153. [CrossRef]

16. Kong, F.D.; Huang, X.L.; Ma, Q.Y.; Xie, Q.Y.; Wang, P.; Chen, P.W.; Zhou, L.M.; Yuan, J.Z.; Dai, H.F.; Luo, D.Q.; Zhao, Y.X. Helvolic acid derivatives with antibacterial activities against *Streptococcus agalactiae* from the marine-derived fungus *Aspergillus fumigatus* HNMF0047. *J. Nat. Prod.* **2018**, *81*, 1869–1876. [CrossRef]

17. Lee, S.Y.; Kinoshita, H.; Ihara, F.; Igarashi, Y.; Nihira, T. Identification of novel derivative of helvolic acid from *Metarhizium anisopliae* grown in medium with insect component. *J. Biosci. Bioeng.* **2008**, *105*, 476–480. [CrossRef]

18. Yamada, T.; Ohshima, M.; Yuasa, K.; Kikuchi, T.; Tanaka, R. Assignment of the CD cotton effect to the chiral center in pseurotins, and the stereochemical revision of Pseurotin A_2. *Mar. Drugs* **2016**, *14*, 74. [CrossRef]

19. Wang, Q.; Song, F.; Xiao, X.; Huang, P.; Li, L.; Monte, A.; Abdel-Mageed, W.M.; Wang, J.; Guo, H.; He, W.; et al. Abyssomicins from the South China Sea deep-sea sediment *Verrucosispora* sp.: Natural thioether Michael addition adducts as antitubercular prodrugs. *Angew. Chem. Int. Ed. Engl.* **2013**, *52*, 1231–1234. [CrossRef]

20. Wenke, J.; Anke, H.; Sterner, O. Pseurotin A and 8-*O*-demethylpseurotin A from *Aspergillus fumigatus* and their inhibitory activities on chitin synthase. *Biosci. Biotech. Bioch.* **1993**, *57*, 961–964. [CrossRef]

21. Wang, F.Z.; Li, D.H.; Zhu, T.J.; Zhang, M.; Gu, Q.Q. Pseurotin A_1 and A_2, two new 1-oxa-7-azaspiro[4.4] non-2-ene-4,6-diones from the holothurian-derived fungus *Aspergillus fumigatus* WFZ-25. *Can. J. Chem.* **2011**, *89*, 72–76. [CrossRef]

22. Breitenstein, W.; Chexal, K.K.; Mohr, P.; Tamm, C. Pseurotin B, C, D, and E. further new metabolites of *Pseudeurotium ovalis* STOLK. *Helv. Chim. Acta.* **1981**, *64*, 379–388. [CrossRef]

23. Wink, J.; Grabley, S.; Gareis, M.; Thiericke, R.; Kirsch, R. Pseurotin F1/F2, New Metabolites from *Aspergillus fumigatus*, Process for Their Preparation and Their Use as Apomorphine Antagonists. European Patent Application EP546474, 1993.

24. Zhang, M.; Wang, W.L.; Fang, Y.C.; Zhu, T.J.; Gu, Q.Q.; Zhu, W.M. Cytotoxic alkaloids and antibiotic nordammarane triterpenoids from the marine-derived fungus *Aspergillus sydowi*. *J. Nat. Prod.* **2008**, *71*, 985–989. [CrossRef] [PubMed]

25. Ando, O.; Satake, H.; Nakajima, M.; Sato, A.; Nakamura, T.; Kinoshita, T.; Furuya, K.; Haneishi, T. Synerazol, a new antifungal antibiotic. *J. Antibiot. (Tokyo)* **1991**, *44*, 382–389. [CrossRef]

26. Asami, Y.; Kakeya, H.; Onose, R.; Yoshida, A.; Matsuzaki, H.; Osada, H. Azaspirene: A novel angiogenesis inhibitor containing a 1-oxa-7-azaspiro[4.4]non-2-ene-4,6-dione skeleton produced by the fungus *Neosartorya* sp. *Org. Lett.* **2002**, *4*, 2845–2848. [CrossRef] [PubMed]

27. Ren, H.; Liu, R.; Chen, L.; Zhu, T.J.; Zhu, W.M.; Gu, Q.Q. Two new hetero-spirocyclic gamma-Lactam derivatives from marine dediment-derived fungus *Aspergillus sydowi* D2-6. *Arch. Pharm. Res.* **2010**, *33*, 499–502. [CrossRef] [PubMed]

28. Yamada, T.; Imai, E.; Nakatuji, K.; Numata, A.; Tanaka, R. Cephalimysin A, a potent cytotoxic metabolite from an *Aspergillus* species separated from a marine fish. *Tetrahedron Lett.* **2007**, *48*, 6294–6296. [CrossRef]

29. Yamada, T.; Kajimoto, T.; Kikuchi, T.; Tanaka, R. Elucidation of the relationship between CD Cotton effects and the absolute configuration of sixteen stereoisomers of spiroheterocyclic-lactams. *Mar. Drugs* **2018**, *16*, 223. [CrossRef]

Communication

Geospallins A–C: New Thiodiketopiperazines with Inhibitory Activity against Angiotensin-Converting Enzyme from a Deep-Sea-Derived Fungus *Geosmithia pallida* FS140

Zhang-Hua Sun [1,2,†], Jiangyong Gu [3,†], Wei Ye [1,†], Liang-Xi Wen [4], Qi-Bin Lin [4], Sai-Ni Li [1], Yu-Chan Chen [1], Hao-Hua Li [1] and Wei-Min Zhang [1,*]

[1] State Key Laboratory of Applied Microbiology Southern China, Guangdong Provincial Key Laboratory of Microbial Culture Collection and Application, Guangdong Open Laboratory of Applied Microbiology, Guangdong Institute of Microbiology, Guangzhou 510070, China; sysuszh@126.com (Z.-H.S.); yewei@gdim.cn (W.Y.); lisn@gdim.cn (S.-N.L.); yuchan2006@126.com (Y.-C.C.); hhli100@126.com (H.-H.L.)
[2] Guangdong Provincial Key Laboratory of New Drug, Development and Research of Chinese Medicine, Mathematical Engineering Academy of Chinese Medicine, Guangzhou University of Chinese Medicine, Guangzhou 510006, China
[3] The Second Institute of Clinical Medicine, Guangzhou University of Chinese Medicine, Guangzhou 510006, China; gujy@gzucm.edu.cn
[4] School of Pharmaceutical Sciences, Guangzhou University of Chinese Medicine, Guangzhou 510006, China; wen_en_liangxi@163.com (L.-X.W.); 15622154060@163.com (Q.-B.L.)
* Correspondence: wmzhang@gdim.cn; Tel.: +86-20-8768-8309
† These authors contributed equally to this work.

Received: 23 October 2018; Accepted: 19 November 2018; Published: 23 November 2018

Abstract: Three new thiodiketopiperazines, geospallins A–C (**1–3**), together with nine known analogues (**4–12**), were isolated from the culture of the deep-sea sediment-derived fungus *Geosmithia pallida* FS140. Among them, geospallins A and B (**1** and **2**) represent rare examples of thiodiketopiperazines featuring an S-methyl group at C-10 and a tertiary hydroxyl group at C-11. Their structures were determined by high-resolution electrospray mass spectrometry (HRESIMS), spectroscopic analyses, and electronic circular dichroism (ECD) calculations. Their angiotensin-converting enzyme (ACE) inhibitory activity was reported, and geospallins A–C (**1–3**) showed inhibitory activity with IC$_{50}$ values of 29–35 µM.

Keywords: thiodiketopiperazines; *Geosmithia pallida*; deep-sea-derived fungus

1. Introduction

2,5-Diketopiperazines are cyclodipeptides obtained by the condensation of two α-amino acids. This subunit is often found alone or embedded in larger, more complex architectures in a variety of natural products from fungi, bacteria, the plant kingdom, and mammals [1]. They are not only a class of naturally occurring privileged structures that have the ability to bind to a wide range of receptors, but they also have several characteristics that make them attractive scaffolds for drug discovery [2,3]. Thiodiketopiperazines, such as gliotoxins, chetoseminudins, luteoalbusins, chetracins, and apoaranotins, are common across the microbial world and have highly diverse organic structures characterized by sulfur-containing functional groups. They also exhibit various pharmacological activities, such as antifungal, antibacterial, and cytotoxic activities [4]. The angiotensin-converting enzyme (ACE) is an important target and has broad effects in different systems, and ACE inhibitors were originally developed for the treatment of congestive heart failure, diabetic kidney disease, and hypertension management [5]. ACE cleaves many peptides besides angiotensin I and thereby affects diverse physiological functions, including renal development and male reproduction. In addition, ACE has a role in both innate and adaptive responses by modulating macrophage and neutrophil function—effects that are magnified when these cells overexpress ACE. Macrophages that overexpress ACE are more effective against tumors and infections [5]. Sulfur-containing metabolites are crucial for the inhibitory activity against ACE, which catalyzes the reaction from angiotensin I to angiotensin II in the renin–angiotensin system and plays a major role in hypertension [6].

Recently, we initiated the investigation of microorganisms derived from deep-sea sediments, aiming at discovering new metabolites with potent bioactivity [7–9]. As part of the program, the fungus *Geosmithia pallida* FS140 was isolated from a sediment collected at a depth of 2403 m in the South China Sea (19°28.581′ N, 115°27.251′ E). Chemical investigation of the fermentation broth led to the isolation of a series of diketopiperazines including three new thiodiketopiperazines, named geospallins A–C (**1–3**), as well as nine known analogues (**4–12**). Geospallins A and B (**1** and **2**) represent rare examples of thiodiketopiperazines featuring an S-methyl group at C-10 and a tertiary hydroxyl group at C-11. Biological tests verified that compounds **1−3** were responsible for the inhibition of ACE. Details of the isolation, structure elucidation, and biological activities of diketopiperazines **1–12** are presented below.

2. Results

The fermentation broth of the deep-sea-derived fungus *G. pallida* FS140 was extracted with EtOAc and then concentrated under reduced pressure to give an extract. The EtOAc extract was subjected to a series of solvent/solvent partitioning steps to afford compounds **1–12** (Figure 1). Three new structures were identified by the combination of spectroscopic analysis, high-resolution electrospray mass spectrometry (HRESIMS), and electronic circular dichroism (ECD) calculation, while twelve known analogues were identified as bisdethiobis (methylthio)gliotoxin (**4**) [10,11], 6-acetylbis(methylthio)gliotoxin (**5**) [9,12], 6-deoxy-5a,6-didehydrogliotoxin (**6**) [13], 5a,6-didehydrogliotoxin (**7**) [14], 6-(phenylmethyl)-(3*R*,6*R*)-2,5-piperazinedione (**8**) [15], 3-(hydroxymethyl)-3,6-bis(methylthio)-6-(phenylmethyl)-(3*R*,6*R*)-2,5-piperazinedione (**9**) [15], 3-(hydroxymethyl)-6-(methoxyl)-6-(phenylmethyl)-(3*R*,6*R*)-2,5-piperazinedione (**10**) [16], 5a,6-anhydrobisdethiobis(methylthio)gliotoxin (**11**) [17], and bisdethiobis (methylthio)gliotoxin (**12**) [12] by comparison of their spectroscopic data with those in the literature.

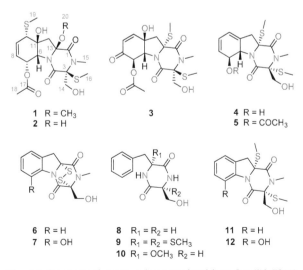

Figure 1. Structures of compounds **1–12** isolated from *G. pallida* FS140.

2.1. Identification of New Compounds

Compound **1**, a colorless oil, had the molecular formula of $C_{18}H_{26}N_2O_7S_2$, as determined by HRESIMS (m/z 469.1063 [M + Na]$^+$, calcd for 469.1074), corresponding to seven degrees of unsaturation. The ^{1}H NMR spectrum revealed the presence of five methyl singlets (δ_H 3.15, 2.99, 2.26, 2.05, and 1.76), a disubstituted double bond (δ_H 6.00 (2H, m)), and a series of aliphatic methylene or methine multiplets (5.60 (1H, s), 5.52 (1H, s), 5.48 (1H, d, J = 5.6 Hz), 4.23 (1H, d, J = 1.8 Hz), 3.93 (1H, dd, J = 11.7, 6.7 Hz), 3.68 (1H, m), 3.66 (1H, m), 2.54 (1H, m), and 1.86 (1H, d, J = 14.2 Hz)). The ^{13}C NMR, in combination with HSQC experiments, resolved 18 carbon resonances attributed to three carbonyl groups (δ_C 169.4, 167.6, and 163.5), a disubstituted double bond (δ_C 137.2 and 127.4), three sp^3 quaternary carbons (δ_C 90.6, 81.0, and 76.8), three sp^3 methines (δ_C 70.5, 65.2, and 51.8), two sp^3 methylene (δ_C 62.8 and 42.3), a methoxyl group (δ_C 52.1), an N-methyl group (δ_C 28.6), and three methyl groups (δ_C 20.6, 16.1, and 11.0) (Table 1). The 1D NMR data of **1** showed resonance characteristics of a thiodiketopiperazine framework similar to that of **5**, except for the absence of a double bond and presence of a methoxy group. In comparison with **5**, the ^{13}C NMR spectroscopic data for **1** differed significantly from C-10 to C-13, with the distinctly upfield-shifted carbon at C-13 (δ_C 90.6 in **1** and δ_C 75.2 in **5**) and the downfield-shifted at C-10 and C-11 (δ_C 51.8 and 81.0 in **1** and δ_C 120.5 and 135.7 in **5**, respectively). Detailed 2D NMR analyses (^{1}H–^{1}H COSY, HSQC, and HMBC) allowed the establishment of the gross structure of **1** as depicted in Figure 2. HMBC correlations from H-7 (δ_H 5.48) and H$_3$-18 (δ_H 2.05) to C-17 (δ_C 169.4), from H$_3$-19 (δ_H 2.26) to C-10 (δ_C 51.8), and from OH-11 (δ_H 5.60) to C-6/C-10/C-11/C-12 (δ_C 70.5, 51.8, 81.0, and 42.3, respectively), and ^{1}H–^{1}H COSY correlations of H-6/H-7/H-8/H-9/H-10 confirmed the presence of fragment A (Figure 2). Fragment B was very similar to **5**, except for the distinct downfield-shifted of the O-methyl group when compared with the S-methyl group (δ_C 52.1 in **1** and δ_C 15.1 in **5**). The methyl group was assigned to C-13 by the HMBC correlations from the methyl group (δ_H 3.15) to the severely downfield-shifted C-13 (90.6 ppm in **1** and 75.2 ppm in **5**, respectively).

The relative configuration of **1** was established on the basis of the interpretation of the NOESY data and ^{1}H–^{1}H coupling constants (Figure 2 and Table 1). The strong NOE interactions of H-10/OH-11, OH-11/H-6, H-6/H-7, OH-11/H-12β, and H-12β/H$_3$-20 indicated that H-10, OH-11, H-6, H-7, and CH$_3$-20 occupied the axial bonds of the cyclohexane-ring portion in a chair conformation and were arbitrarily assigned β-orientations. Additionally, the axial orientation of both H-6 and H-7 were in good accordance with their small coupling constant of 1.8 Hz.

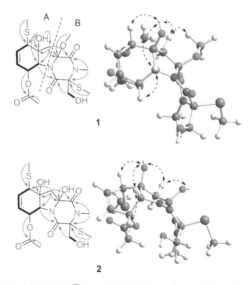

Figure 2. Key COSY (—), HMBC (⌒↘), and NOESY (◄- - -►) correlations for compounds **1** and **2**.

Table 1. ^{1}H (400 MHz) and ^{13}C (100 MHz) NMR data for compounds **1–3** (in DMSO-d_6)[a].

No.	1		2		3	
	^{1}H	^{13}C	^{1}H	^{13}C	^{1}H	^{13}C
1		163.5		166.1		165.0
3		76.8		77.9		70.8
4		167.6		167.6		162.6
6	4.23, d (1.8)	70.5	4.32, d (1.8)	70.5	4.96, d (11.0)	69.8
7	5.48, d (5.6)	65.2	5.32, d (3.2)	66.4	5.80, d (11.0)	75.1
8	6.00, m	127.4	5.96, d (3.2)	127.4		191.6
9	6.00, m	137.2	5.96, d (3.2)	137.2	6.09, d (10.3)	125.4
10	3.66, m	51.8	3.65, s	51.3	6.98, d (10.3)	150.5
11		81.0		81.4		75.1
12	2.54, m / 1.86, d (14.2)	42.3	2.28, d (8.5) / 2.12, d (14.2)	45.6	3.07, d (14.9) / 2.95, d (14.9)	49.3
13		90.6		86.3		69.1
14	3.93, dd (11.7, 6.7) / 3.68, m	62.8	3.81, m / 3.71, dd (10.9, 3.6)	62.2	4.18, dd (11.5, 6.0) / 3.74, dd (11.5, 4.7)	62.6
N-Me	2.99, s	28.6	2.99, s	29.1	2.99, s	28.8
SMe-3	1.76, s	11.0	1.80, d (3.6)	10.6	2.16, s	13.0
OAc-7		169.4		169.4		168.9
	2.05, s	20.6	2.05, s	20.6	2.07, s	20.4
SMe-10	2.26, s	16.1	2.26, s	16.1		
OMe-13	3.15, s	52.1			2.18, s	14.6
11-OH	5.60, s		5.68, s		6.00, s	
13-OH			6.04, brs			
14-OH	5.52, s		6.64, brs		5.34, t (5.4)	

[a] Chemical shifts are in ppm; coupling constant *J* is in Hz.

In order to define the absolute configuration of **1**, the ECD spectrum of (3*R*, 6*R*, 7*R*, 10*S*, 11*R*, 13*R*)-**1**, (3*R*, 6*R*, 7*R*, 10*S*, 11*R*, 13*S*)-**1**, (3*S*, 6*S*, 7*S*, 10*R*, 11*S*, 13*R*)-**1**, and (3*S*, 6*S*, 7*S*, 10*R*, 11*S*, 13*S*)-**1** were calculated by the time-dependent density functional theory (TDDFT) computational method and compared with the experimental spectra of **1** (for details of calculations, see Supplementary Materials). The experimental ECD spectrum of **1** showed an ECD curve with positive Cotton effects around 219 nm (Figure 3a). The calculated ECD spectrum for (3*S*, 6*S*, 7*S*, 10*R*, 11*S*, 13*S*)-**1** showed a similar ECD curve with positive Cotton effects at 220 nm, indicating that **1** had an (3*S*, 6*S*, 7*S*, 10*R*, 11*S*, 13*S*)-configuration. Compound **1** was given the trivial name geospallin A.

The HRESIMS data of **2** exhibited a sodium adduct ion at *m/z* 455.0907 [M + Na]$^+$ (calcd 455.0917), consistent with the molecular formula $C_{17}H_{24}N_2O_7S_2Na$, showing 14 mass units less than that of **1**. The 1D NMR data of **2** were similar to those of **1**, except for the absence of a methoxy group in **2**, indicating **2** was a demethylated derivative of **1**. The additional hydroxy group was located at C-13 by HMBC correlation from the hydroxyl group (δ_H 6.04) to C-13 (δ_C 86.3) (Figure 2). The absolute configuration of **2** was confirmed by using the same methods as described for **1**. The experimental ECD spectrum for **2** showed a similar ECD curve with positive Cotton effects at 220 nm (Figure S46 in Supplementary Materials), indicating that compound **2** has a (3*S*, 6*S*, 7*S*, 10*R*, 11*S*, 13*S*)-configuration. Compound **2** was given the trivial name geospallin B.

Compound **3**, a colorless oil, had the molecular formula of $C_{17}H_{22}N_2O_7S_2$, as determined by HRESIMS (*m/z* 453.0764 [M + Na]$^+$, $C_{17}H_{22}N_2O_7S_2Na$, calcd for 453.0761), corresponding to eight degrees of unsaturation. The ^{1}H and ^{13}C NMR spectra (Table 1) of **3** showed high similarity to those of **5**, except for the presence of an α,β-unsaturated ketone moiety (δ_C 191.6, 125.4, and 150.5) in **3** instead of two double bonds in **5**, indicating that **3** was an oxidative derivative of **5**. This was supported by detailed 2D NMR spectra analyses (Figure 4). The location of the α,β-unsaturated ketone moiety was assigned at C-11 by HMBC correlations from OH-11 (δ_H 6.00) to C-10, C-11, C-6, and C-12 (δ_C 150.5, 75.1, 69.8, and 49.3, respectively); from H-10 (δ_H 6.98) to C-6 and C-8 (δ_C 191.6); and from H-7 (δ_H 5.80) to the carbonyl group (δ_C 191.6, C-8) and acetyl group (δ_C 168.9). The relative configuration of **3** was deduced by NOESY correlations of H-6/H-12β, H-7/H12α, and H-12α/SMe-13.

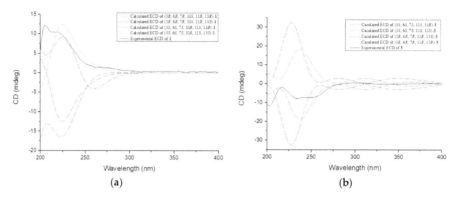

(a) (b)

Figure 3. (**a**) Experimental electronic circular dichroism (ECD) spectra of geospallin A (**1**) in MeOH and calculated ECD spectra of (3*R*, 6*R*, 7*R*, 10*S*, 11*R*, 13*R*)-**1**, (3*R*, 6*R*, 7*R*, 10*S*, 11*R*, 13*S*)-**1**, (3*S*, 6*S*, 7*S*, 10*R*, 11*S*, 13*R*)-**1**, and (3*S*, 6*S*, 7*S*, 10*R*, 11*S*, 13*S*)-**1**; (**b**) Experimental ECD spectra of geospallin C (**3**) in MeOH and calculated ECD spectra of (3*S*, 6*S*, 7*S*, 11*S*, 13*R*)-**3**, (3*S*, 6*S*, 7*S*, 11*S*, 13*S*)-**3**, (3*R*, 6*R*, 7*R*, 11*R*, 13*S*)-**3**, and (3*R*, 6*R*, 7*R*, 11*R*, 13*R*)-**3**. The calculated ECD spectra were computed at the B3LYP/6-311G (2d+p) level.

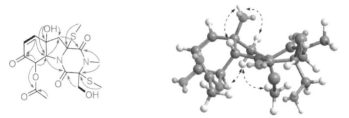

Figure 4. Key ^{1}H-^{1}H COSY (▬), HMBC (⌒◝), and NOESY (◄- - -►) correlations for geospallin C (**3**).

The absolute structure of **3** was also deduced by comparison of the experimental and calculated ECD spectra generated by TDDFT calculations in the Gaussian 16 program (Figure 3b). As illustrated in Figure 3b, the experimentally acquired ECD spectrum for **3** agreed well with the ECD curve computed for (3*S*, 6*S*, 7*S*, 11*S*, 13*S*)-**3**. Compound **3** was given the trivial name geospallin C.

2.2. Angiotensin-Converting Enzyme (ACE) Inhibitory Assay

The twelve thiodiketopiperazines (**1**–**12**) reported in this study were produced in sufficient amounts to allow testing for the inhibition of angiotensin-converting enzyme (Table 2). IC$_{50}$ values above 100 μM were not determined, while compounds **1**−**3** showed inhibitory activity against ACE with an IC$_{50}$ value range of 29−35 μM.

Table 2. IC$_{50}$ values of compounds **1**–**3** against angiotensin-converting enzyme (ACE).

Compounds	IC$_{50}$ (μM)
1	35 ± 5.2
2	31 ± 3.3
3	29 ± 3.3
Captopril	0.041 ± 0.005

2.3. α-Glucosidase Inhibitory Activity Assay

All compounds were evaluated in vitro for α-glucosidase inhibitory activity. However, none of the compounds showed inhibitory activity against α-glucosidase at a concentration of 100 μM.

3. Materials and Methods

3.1. General Experimental Procedures

IR spectra were carried out on a Shimadzu IR Affinity-1 spectrophotometer (Shimadzu Corporation, Kyoto, Japan). UV data was acquired using a Shimadzu UV-2600 spectrophotometer (Shimadzu Corporation, Kyoto, Japan). Optical rotations were obtained on an Anton Paar MCP-500 (Anton Paar, Graz, Austria). Circular dichroism (CD) spectra were recorded on a Jasco 820 spectropolarimeter (Jasco Corporation, Kyoto, Japan). NMR spectra were determined on a Bruker Avance-400 spectrometer (Bruker Corporation, Fremont, CA, USA). ESI-MS spectra were measured on an Agilent Technologies 1290-6430A Triple Quad LC/MS (Agilent Technologies Inc., Santa Clara, CA, USA), and HRESIMS was measured on a Thermo MAT95XP high-resolution mass spectrometer (Thermo Fisher Scientific, Bremen, Germany). A Shimadzu LC-20 AP (Shimadzu Corporation, Kyoto, Japan) equipped with an SPD-M20A Photo-Diode Array (PDA) detector (Shimadzu Corporation, Kyoto, Japan) was used for HPLC analysis. A YMC-pack ODS-A column (250 × 20 mm, 5 μm, 12 nm, YMC CO., Ltd., Kyoto, Japan) was used for preparative HPLC separation. Column chromatography was conducted using a commercial silica gel (SiO_2; 200–300 mesh; Qingdao Haiyang Chemical Co. Ltd., Qingdao, China) and Sephadex LH-20 gel (Amersham Biosciences, Uppsala, Sweden). All solvents were of analytical grade (Guangzhou Chemical Regents Company, Ltd., Guangzhou, China).

3.2. Fungal Material and Identification

The fungal strain FS140 was isolated from a deep-sea sludge in the South China Sea (19°28.581′ N, 115°27.251′ E, depth 2403 m) in September 2011. The isolate was identified as *Geosmithia pallida* FS140 by sequence analysis of the internal transcribed spacer (ITS) region of the ribosomal DNA. The sequence data have been submitted to GenBank (accession no. MK047400), and FS140 has 99% similarity with *Geosmithia pallida* CCF4279 (accession no. KF808303). The strain was deposited in the Guangdong Provincial Key Laboratory of Microbial Culture Collection and Application, Guangdong Institute of Microbiology, Guangzhou, People's Republic of China. Working stocks were prepared on potato dextrose agar (PDA) slants and stored at 4 °C.

3.3. Fermentation, Extraction, and Isolation

Three pieces (0.5 × 0.5 cm^2) of mycelial agar plugs of *G. pallida* FS140 were inoculated into 250 mL of PD medium (potato 200 g/L, glucose 20 g/L, KH_2PO_4 3 g/L, $MgSO_4·7H_2O$ 1.5 g/L, vitamin B_1 10 mg/L, sea salt 15 g/L) in 500 mL Erlenmeyer flasks and incubated for 2 days in a rotary shaker (200 r/m) at 28 °C. The seed cultures (10%) were then aseptically transferred into 500 mL of PD medium in 1000 mL Erlenmeyer flasks and kept shaking (120 r/m) at 28 °C for 7 days. The whole fermentation broth (80 L) was filtered through cheese cloth to separate the supernatant from the mycelia. The supernatant was extracted with EtOAc (4 × 25 L) and evaporated under reduced pressure to give a dark brown oily residue (31 g). The EtOAc-soluble fraction was separated over a column of silica gel and eluted with petroleum ether/EtOAc in a linear gradient (30:1 → 1:1) and followed by $CHCl_3$/MeOH in linear gradient (10:1 → 0:1) to give 18 fractions (F1–F18). F7 was separated on a preparative reversed-phase (RP) HPLC system equipped with a C-18 column (YMC*GEL ODS-A, 120A S-5 μm, 250 × 20 mm, MeOH/H_2O, 0.1:0.9 → 1.0:0, 10 mL/min) to give 25 fractions (F7a–F7y). F7c was chromatographed over a Sephadex LH-20 column eluted with $CHCl_3$/MeOH (1:1, *v/v*), then further separated by preparative RP HPLC on the ODS-A column (MeCN/H_2O, 40:60, 10 mL/min) to yield **2** (4.5 mg, t_R = 16.0 min), **9** (23.0 mg, t_R = 25.0 min), **10** (6.1 mg, t_R = 31.0 min), and **8** (7.6 mg, t_R = 34 min), sequentially. F7f was successfully separated by preparative RP HPLC on the

ODS-A column (MeCN/H$_2$O, 45:55, 10 mL/min) to afford **1** (5.9 mg, t_R = 12.1 min) and **3** (6.8 mg, t_R = 13.4 min). F7k was separated on a preparative RP HPLC on the ODS-A column (MeCN/H$_2$O, 50:50, 10 mL/min) to yield **4** (427 mg), and F7m was purified by column chromatography on a Sephadex LH-20 (CHCl$_3$/MeOH, 1:1, *v*/*v*) to afford **5** (312 mg). F9 was purified by preparative HPLC on the ODS-A column (MeCN/H$_2$O, 60:40, 10 mL/min) to give **7** (28 mg, t_R = 14.2 min) and **12** (16.8 mg, t_R = 19.3 min), while F10 was subjected to a Sephadex LH-20 column elution with CHCl$_3$/MeOH (1:1), then further separated by preparative HPLC on the ODS-A column (MeCN/H$_2$O, 65:35, 10 mL/min) to obtain **6** (39 mg, t_R = 9.6 min) and **11** (18.1 mg, t_R = 11.8 min), successively.

Geospallin A (**1**): colorless oil; $[\alpha]_D^{25}$ +87.7 (*c* 0.1, MeOH); CD (MeOH, *c* 0.001 mg/mL) 219 nm ($\Delta\varepsilon$ + 3.16); IR ν_{max} 3443, 2924, 1739, 1662, 1372, 1237, 1095, 1022 cm^{-1}; ^{1}H and ^{13}C NMR data, see Table 1; HRESIMS *m/z* 469.1063 ([M + Na]$^+$, calcd for 469.1074).

Geospallin B (**2**): colorless oil; $[\alpha]_D^{25}$ +84.9 (*c* 0.1, MeOH); CD (MeOH, *c* 0.001 mg/mL) 225 nm ($\Delta\varepsilon$ + 3.21); IR ν_{max} 3315, 3056, 2923, 1738, 1667, 1422, 1373, 1265, 1236 cm^{-1}; ^{1}H and ^{13}C NMR data, see Table 1; HRESIMS *m/z* 455.0907 ([M + Na]$^+$, calcd for 455.0917).

Geospallin C (**3**): colorless oil; $[\alpha]_D^{25}$ −28.2 (*c* 0.1, MeOH); CD (MeOH, *c* 0.001 mg/mL) 202 ($\Delta\varepsilon$ − 3.52), 234 ($\Delta\varepsilon$ − 2.36) nm; UV (MeOH) λ_{max} (log ε) 210 (4.26) nm; IR ν_{max} 3407, 2925, 1749, 1703, 1654, 1415, 1376, 1222, 1085, 1038 cm^{-1}; ^{1}H and ^{13}C NMR data, see Table 1; HRESIMS *m/z* 453.0764 ([M + Na]$^+$, calcd for 453.0761).

3.4. Quantum Chemical ECD Calculation

The quantum chemical ECD calculation methods were used to establish the absolute configurations of compounds **1**–**3**. The 3D structures were generated by Discover Studio 2.5. The conformational search was performed by the Conformer Searching module of Open Babel 2.4.1 using a genetic algorithm and the MMFF94 molecular mechanics force field. The geometry optimizations were then performed by using density functional theory (DFT) at the b3lyp/6-311+g(2d,p) level. These stable conformers, which had no imaginary frequency, were subsequently submitted to ECD calculations by TDDFT calculations at the b3lyp/6-311+g(2d,p) level. The solvent effects were taken into account by the integral equation formalism polarizable continuum model (IEFPCM, methanol). All calculations were performed with the Gaussian 16 A.03 program [18]. The calculated spectra were drawn using SpecDis software with a UV shift to the ECD spectra.

3.5. Angiotensin-Converting Enzyme (ACE) Inhibitory Assay

ACE inhibitory activity was determined by a previously reported method [19]. In brief, 20 µL thiodiketopiperazines dissolved in DMSO with different concentrations were added to 120 µL N-hippuryl–His–Leu substrates, then preheated in water for 3–5 min; next, 10 µL ACE enzymatic solution was added and mixtures were incubated at 37 °C for 60 min, and 150 µL 1 M HCl was added to stop the reaction. The mixture was loaded for HPLC with a flow rate of 0.5 mL/min by 60% methanol, and the absorbance at 228 nm was detected. To serve as a blank, 10 µL pH 8.3 boric acid replaced the thiodiketopiperazines. Hippuric acid solutions at 10, 20, 40, 60, 80, and 100 µg/mL were prepared using 10 µL pH 8.3 boric acid. One enzymatic unit was defined as the amount giving the production of 1 µM hippuric acid by the catalyzation of the substrate N-hippuryl–His–Leu at 37 °C in 1 min. Captopril was used as a positive control. Physiological saline with a concentration of 0.9% (*w*/*w*) was used as a negative control in our experiment, which exhibited no inhibitory activity towards ACE. The IC$_{50}$ values of thiodiketopiperazines were calculated after the ACE inhibitory experiments were conducted, using different concentrations of thiodiketopiperazines.

3.6. α-Glucosidase Inhibitory Activity Assay

An assay of α-glucosidase inhibitory activity was performed as previously described [20].

4. Conclusions

In this study, twelve diketopiperazines, including three new thiodiketopiperazines, were isolated from the deep-sea-derived fungus *Geosmithia pallida*. All the stereochemical configurations of the new compounds, including their absolute configurations, were established. Geospallins A and B (**1** and **2**) represent rare examples of thiodiketopiperazines featuring an S-methyl group at C-10 and a tertiary hydroxyl group at C-11, and their *S*-configuration at both C-7 and C-13 is the first report of such. These thiodiketopiperazines were examined for their angiotensin-converting enzyme inhibitory assay, and geospallins A–C (**1–3**) showed inhibitory activity, with IC_{50} values of 29–35 μM.

Supplementary Materials: The following are available online at http://www.mdpi.com/1660-3397/16/12/464/s1, The ^{1}H- and ^{13}C-NMR data of **1–12** and the HRESIMS and 2D-NMR spectra of compounds **1–3** (Figures S1–S47).

Author Contributions: Z.-H.S. fractionated the extract, isolated and elucidated structures, and wrote the paper; L.-X.W. and Q.-B.L. fractionated the extract and isolated a proportion of the compounds; J.G., performed the calculated ECD spectra experiment; S.-N.L. isolated and identified the fungal strain FS140; W.Y., Y.-C.C., and H.-H.L. performed the bioassays; W.-M.Z. designed and coordinated the study and reviewed the manuscript.

Funding: This work was supported financially by the Science and Technology Program of Guangzhou, China (201607020018), the Team Project of the Natural Science Foundation of Guangdong Province (2016A030312014), the National Natural Science Foundation of China (31272087), and the Guangdong Provincial Innovative Development of Marine Economy Regional Demonstration Projects (GD2012-D01-002).

Acknowledgments: The calculations were performed on the high-performance computer cluster of the Guangdong Provincial Hospital of Traditional Chinese Medicine.

Conflicts of Interest: The authors declare no conflict of interest.

References

1.	Borthwick, A.D. 2,5-Diketopiperazines: Synthesis, reactions, medicinal chemistry, and bioactive natural products. *Chem. Rev.* **2012**, *112*, 3641–3716. [CrossRef]

2.	Blunt, J.W.; Copp, B.R.; Keyzers, R.A.; Munro, M.H.G.; Prinsep, M.R. Marine natural products. *Nat. Prod. Rep.* **2016**, *36*, 382–431. [CrossRef] [PubMed]

3.	Blunt, J.W.; Copp, B.R.; Munro, M.H.G.; Northcote, P.T.; Prinsep, M.R. Marine natural products. *Nat. Prod. Rep.* **2011**, *28*, 196–268. [CrossRef] [PubMed]

4.	Huang, R.; Zhou, X.; Xu, T.; Yang, X.; Liu, Y. Diketopiperazines from marine organisms. *Chem. Biodivers.* **2010**, *7*, 2809–2829. [CrossRef] [PubMed]

5.	Bernstein, K.E.; Khan, Z.; Giani, J.F.; Cao, D.Y.; Bernstein, E.A.; Shen, X.Z. Angiotensin-converting enzyme in innate and adaptive immunity. *Nat. Rev. Nephrol.* **2018**, *14*, 325–336. [CrossRef] [PubMed]

6.	Nakabayashi, R.; Yang, Z.; Nishizawa, T.; Mori, T.; Saito, K. Top-down targeted metabolomics teveals a sulfur-containing metabolite with inhibitory activity against angiotensin-converting enzyme in *Asparagus officinalis*. *J. Nat. Prod.* **2015**, *78*, 1179–1183. [CrossRef] [PubMed]

7.	Xu, J.L.; Liu, H.X.; Chen, Y.C.; Tan, H.B.; Guo, H.; Xu, L.Q.; Li, S.N.; Huang, Z.L.; Li, H.H.; Gao, X.X.; Zhang, W.M. Highly substituted benzophenone aldehydes and eremophilane derivatives from the deep-sea derived fungus *Phomopsis lithocarpus* FS508. *Mar. Drugs* **2018**, *16*, 329. [CrossRef] [PubMed]

8.	Xu, J.; Tan, H.; Chen, Y.; Li, S.; Huang, Z.; Guo, H.; Li, H.; Gao, X.; Liu, H.; Zhang, M.W. Lithocarpins A–D: Four tenellone-macrolide conjugated [4 + 2] hetero-adducts from the deep-sea derived fungus *Phomopsis lithocarpus* FS508. *Org. Chem. Front.* **2018**, *5*, 1792–1797. [CrossRef]

9.	Fan, Z.; Sun, Z.-H.; Liu, Z.; Chen, Y.-C.; Liu, H.-X.; Li, H.-H.; Zhang, W.M. Dichotocejpins A–C: New diketopiperazines from a deep-sea-derived fungus *Dichotomomyces cejpii* FS110. *Mar. Drugs* **2016**, *14*, 164. [CrossRef] [PubMed]

10.	Gordon, W.K.; David, J.R.; Mark, A.S.; Ratnaker, R.T. Biosynthesis of bisdethiobis(methylthio)gliotoxin, a new metabolite of *Gliocladium deliquescens*. *J. Chem. Soc. Perk. Trans. 1* **1980**, *1*, 119–121. [CrossRef]

11.	Afiyatullov, S.S.; Kalinovskii, A.I.; Pivkin, M.V.; Dmitrenok, P.S.; Kuznetsova, T.A. Alkaloids from the marine isolate of the fungus *Aspergillus fumigatus*. *Chem. Nat. Compd.* **2005**, *41*, 236–238. [CrossRef]

12. Liang, W.L.; Le, X.; Li, H.J.; Yang, X.L.; Chen, J.X.; Xu, J.; Liu, H.L.; Wang, L.Y.; Wang, K.T.; Hu, K.C.; Yang, D.P.; Lan, W.J. Exploring the chemodiversity and biological activities of the secondary metabolites from the marine fungus *Neosartorya pseudofischeri*. *Mar. Drugs* **2014**, *12*, 5657–5676. [CrossRef] [PubMed]
13. Sun, Y.; Takada, K.; Takemoto, Y.; Yoshida, M.; Nogi, Y.; Okada, S.; Matsunaga, S. Gliotoxin analogues from a marine-derived fungus, *Penicillium* sp., and their cytotoxic and histone methyltransferase inhibitory activities. *J. Nat. Prod.* **2012**, *75*, 111–114. [CrossRef] [PubMed]
14. Forseth, R.R.; Fox, E.M.; Chung, D.; Howlett, B.J.; Keller, N.P.; Schroeder, F.C. Identification of cryptic products of the gliotoxin gene cluster using NMR-based comparative metabolomics and a model for gliotoxin biosynthesis. *J. Am. Chem. Soc.* **2011**, *133*, 9678–9681. [CrossRef] [PubMed]
15. Isaka, M.; Palasarn, S.; Rachtawee, P.; Vimuttipong, S.; Kongsaeree, P. Unique diketopiperazine dimers from the insect pathogenic fungus *Verticillium hemipterigenum* BCC 1449. *Org. Lett.* **2005**, *7*, 2257–2260. [CrossRef] [PubMed]
16. Zhao, W.Y.; Zhu, T.J.; Fan, G.T.; Liu, H.B.; Fang, Y.C.; Gu, Q.Q.; Zhu, W.M. Three new dioxopiperazine metabolites from a marine-derived fungus *Aspergillus fumigatus* Fres. *Nat. Prod. Res.* **2010**, *24*, 953–957. [CrossRef] [PubMed]
17. Harms, H.; Orlikova, B.; Ji, S.; Nesaei-Mosaferan, D.; König, G.M.; Diederich, M. Epipolythiodiketopiperazines from the marine derived fungus *Dichotomomyces cejpii* with NF-κB inhibitory potential. *Mar. Drugs* **2015**, *13*, 4949–4966. [CrossRef] [PubMed]
18. Frisch, M.J.; Trucks, G.W.; Schlegel, H.B.; Scuseria, G.E.; Robb, M.A.; Cheeseman, J.R.; Scalmani, G.; Barone, V.; Mennucci, B.; Petersson, G.A.; et al. Gaussian 16W, Revision A.03. Gaussian, Inc.: Wallingford, CT, USA, 2016.
19. Centeno, J.M.; Burguete, M.C.; Castelló-Ruiz, M.; Enrique, M.; Vallés, S.; Salom, J.B.; Torregrosa, G.; Marcos, J.F.; Alborch, E.; Manzanares, P. Lactoferricin-related peptides with inhibitory effects on ACE-dependent vasoconstriction. *J. Agric. Food. Chem.* **2006**, *54*, 5323–5329. [CrossRef] [PubMed]
20. Feng, J.; Yang, X.W.; Wang, R.F. Bio-assay guided isolation and identification of α-glucosidase inhibitors from the leaves of *Aquilaria sinensis*. *Phytochemistry* **2011**, *72*, 242–247. [CrossRef] [PubMed]

Article

New Ophiobolin Derivatives from the Marine Fungus *Aspergillus flocculosus* and Their Cytotoxicities against Cancer Cells

Byeoung-Kyu Choi [1,2], Phan Thi Hoai Trinh [3,4], Hwa-Sun Lee [2], Byeong-Woo Choi [2], Jong Soon Kang [5], Ngo Thi Duy Ngoc [3], Tran Thi Thanh Van [3,4] and Hee Jae Shin [1,2,*]

[1] Department of Marine Biotechnology, University of Science and Technology (UST), 217 Gajungro, Yuseong-gu, Daejeon 34113, Korea; choibk4404@kiost.ac
[2] Marine Natural Products Chemistry Laboratory, Korea Institute of Ocean Science and Technology, 385 Haeyang-ro, Yeongdo-gu, Busan 49111, Korea; hwasunlee@kiost.ac (H.-S.L.); choibw0924@gmail.com (B.-W.C.)
[3] Nhatrang Institute of Technology Research and Application, Vietnam Academy of Science and Technology, 02 Hung Vuong, Nha Trang 650000, Vietnam; phanhoaitrinh84@gmail.com (P.T.H.T.); ngoduyngoc@nitra.vast.vn (N.T.D.N.); tranthanhvan@nitra.vast.vn (T.T.T.V.)
[4] Graduate University of Science and Technology, Vietnam Academy of Science and Technology, 18 Hoang Quoc Viet, Cau Giay, Ha Noi 100000, Vietnam
[5] Laboratory Animal Resource Center, Korea Research Institute of Bioscience and Biotechnology, 30 Yeongudanjiro, Cheongju 28116, Korea; kanjon@kribb.re.kr
* Correspondence: shinhj@kiost.ac.kr; Tel.: +82-51-664-3341; Fax: +82-51-664-3340

Received: 27 May 2019; Accepted: 6 June 2019; Published: 11 June 2019

Abstract: Five new sesterterpenes, 14,15-dehydro-6-*epi*-ophiobolin K (**1**), 14,15-dehydro-ophiobolin K (**2**), 14,15-dehydro-6-*epi*-ophiobolin G (**3**), 14,15-dehydro-ophiobolin G (**4**) and 14,15-dehydro-(*Z*)-14-ophiobolin G (**5**), together with four known ophiobolins (**6–9**) were isolated from the marine fungus *Aspergillus flocculosus* derived from the seaweed *Padina* sp. collected in Vietnam. The five new ophiobolins were first isolated as ophiobolin derivatives consisting of a fully unsaturated side chain. Their structures were elucidated via spectroscopic methods including 1D, 2D NMR and HR-ESIMS. The absolute configurations were determined by the comparison of chemical shifts and optical rotation values with those of known ophiobolins. All compounds (**1–9**) were then evaluated for their cytotoxicity against six cancer cell lines, HCT-15, NUGC-3, NCI-H23, ACHN, PC-3 and MDA-MB-231. All the compounds showed potent cytotoxicity with GI_{50} values ranging from 0.14 to 2.01 μM.

Keywords: ophiobolins; marine fungus; *Aspergillus flocculosus*; anti-proliferation

1. Introduction

The marine environment is an enormous reservoir of novel sources of biologically active metabolites, many of which display unique structural skeletons that can be used as lead structures for the development of new drugs [1,2]. To adapt and live in an environment that is significantly different from terrestrial organisms, marine organisms frequently produce structurally unique chemical compounds [3,4]. Specifically, secondary metabolites from marine microorganisms are recognized as a novel chemical source for drug discovery and development. Among marine-derived microbes, marine fungi produce a wide range of promising biologically active compounds [5]. Numerous novel compounds from marine fungi have displayed a wide range of bioactivities such as antiviral, antibacterial, anticancer, antiplasmodial and anti-inflammatory [6–8]. In the marine context, *Aspergillus* and *Penicillium* represent the best studied fungal genera as depicted in marine contexts [9,10]. The genus *Aspergillus* is known as a major contributor of pharmacologically bioactive compounds, including anticancer asperazine, antibacterial varixanthone and antifungal amphotericin B [11,12].

Ophiobolins are a group of sesterpenoids with an unusual tricyclic 5-8-5 ring system. They show a broad range of inhibitory activities against nematodes, fungi, bacteria and cytotoxic activity against cancer cells [13,14]. They are produced by the fungal genus *Bipolaris, Aspergillus, Sarocladium* and *Drechslera* [15]. The first ophiobolin, ophiobolin A, was isolated from *Biolaris* spp. and displays inhibitory activity against calmodulin-activated cyclic nucleotide phosphodiesterase [16]. These findings made the compound a useful calmodulin probe for research purposes and implied an application in anti–cancer therapy [17]. Interestingly, more than half of the 49 ophiobolins identified between 1999 and 2016 exhibit cytotoxic activities against human cancer cell lines [18]. Although their biological properties have been well exploited in recent years, their structure-activity relationship remains unestablished [18]. Consequently, this study focused on the discovery of bioactive natural products from marine fungi. During our ongoing investigation for new bioactive compounds from marine microorganisms, the fungal 168ST-16.1 strain was isolated from the seaweed *Padina* sp. collected at Da Nang, Vietnam, and, based on its 28S rRNA gene sequence, it was identified as *Aspergillus flocculosus*. Subsequent chemical investigations on an EtOAc extract of the fungal culture broth using reversed-phase HPLC led to the isolation of the five new ophiobolins, named, 14,15-dehydro-6-*epi*-ophiobolins K and G (**1** and **3**), 14,15-dehydro-ophiobolins K and G (**2** and **4**) and 14,15-dehydro-(Z)-14-ophiobolin G (**5**), together with four known ophiobolins, 6-*epi*-ophiobolins C and N (**6** and **8**) and ophiobolins C and N (**7** and **9**) [19–21] (Figure 1). Herein, details of the structure elucidation and biological activity of these compounds are described.

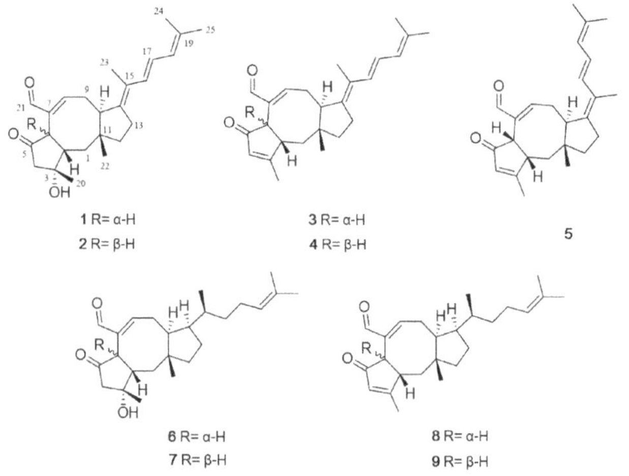

Figure 1. Structures of **1**–**9** isolated from *Aspergillus flocculosus*.

2. Results and Discussion

Compound **1** was obtained as an amorphous powder. The molecular formula of **1** was determined to be $C_{25}H_{34}O_3$ based on HRESIMS. The ^{1}H NMR spectroscopic data of **1** displayed resonances for an aldehyde proton (δ_H 9.23), four olefinic protons (δ_H 6.84, 6.42, 6.38 and 5.93), five methylene protons (δ_H 3.15, 2.47, δ_H 2.94, 2.22, δ_H 2.59, 2.25, δ_H 1.84, 1.63 and δ_H 1.63, 1.50), three methine protons (δ_H 3.33, 3.19 and 2.19) and five methyl groups (δ_H 1.83, 1.81, 1.79, 1.47 and 0.94) (Table 1). The combination of ^{13}C NMR and HSQC spectra revealed the presence of 25 carbon resonances, including one ketone (δ_C 216.6), one aldehyde carbon (δ_C 194.0), four olefinic carbons, five methylene, three methine, five methyl and six quaternary carbons (δ_C 146.8, 142.1, 134.9, 124.9, 76.6 and 43.6) (Table 2). Spin systems and their partial structures were confirmed and assembled by combined analysis of COSY and HMBC correlations (Figure 2). Three spin systems, H_2-1/H-2/H-6, H-8/H_2-9/H-10 and H_2-12/H_2-13, and HMBC correlations from H_3-22 (δ_H 0.94) to C-1 (δ_C 41.4), C-10 (δ_C 47.8), C-11 (δ_C 43.6) and C-12 (δ_C 44.5), and from H-21 (δ_H 9.23) to C-6 (δ_C 48.8), C-7 (δ_C 142.1) and C-8 (δ_C 158.4) confirmed the presence of an eight-membered ring system with an aldehyde group. The five-membered ring with a ketone was also determined by the HMBC correlations from H_3-20 (δ_H 1.47) to C-2 (δ_C 49.6), C-3 (δ_C 76.6) and C-4 (δ_C 55.0) and from H-6 (δ_H 3.33) to C-4 (δ_C 55.0), C-5 (δ_C 216.6), C-7 (δ_C 142.1) and C-21 (δ_C 194.0). Additionally, the HMBC correlations from H_2-13 (δ_H 2.25, 2.59) to C-10 (δ_C 47.8), C-14 (δ_C 146.8) and C-15 (δ_C 124.9) suggested that one additional five-membered ring was connected to the eight-membered ring, which generated a 5-8-5 tricyclic carbon skeleton. The partial structure was closely related to ophiobolin analogs and the ^{1}H and ^{13}C NMR spectra of **1** resembled those of 6-*epi*-ophiobolin C (**6**) except for the presence of two olefinic protons (δ_H 6.38 and 6.42) and two sp^2 quaternary carbons (δ_C 146.8 and 124.9). Finally, the COSY correlation of H-16/H-17/H-18 and the HMBC correlations from H_3-24 (δ_H 1.81) and H_3-25 (δ_H 1.79) to C-18 (δ_C 125.9) and C-19 (δ_C 134.9) and from H_3-23 (δ_H 1.83) to C-14 (δ_C 146.8), C-15 (δ_C 124.9) and C-16 (δ_C 130.6) defined a conjugated side chain connected to C-14 of the tricyclic ring. The planar structure of **1** was elucidated to possess a fully unsaturated side chain. To the best of our knowledge, **1** is the first ophiobolin with three double bonds at the side chain and is named 14,15-dehydro-6-*epi*- ophiobolin K.

Table 1. ^{1}H and ^{13}C NMR data for **1**, **2** and **3** at 600 MHz (δ in ppm, *J* in Hz).

Position	1 [a]	2 [b]	3 [b]	4 [b]	5 [b]
1α	1.63, m	1.54, m	1.22 (t, 13.2)	1.44, m	1.42, m
1β	1.84, m	1.54, m	2.15 (dd, 13.2, 3.6)	2.13 (dd, 15.8, 4.5)	2.13 (dd, 15.8, 4.5)
2	2.19, m	2.57, m	2.72 (d, 12.9)	3.32, overlap	3.31, overlap
4	2.47 (d, 16.6)	2.57 (d, 18.9)	6.01, s	6.09, s	6.08, s
	3.15 (d, 16.6)	2.70 (d, 18.9)			
6	3.33 (d, 10.5)	3.47 (d, 11.8)	3.53 (d, 3.6)	4.29 (d, 7.2)	4.28 (d, 7.2)
8	6.84 (d, 6.7)	7.44 (t, 8.6)	6.91 (dd, 6.2, 2.3)	7.24 (d, 6.7)	7.21 (d, 6.8)
9α	2.94 (d, 20.8)	2.48, m	3.02 (d, 21.3)	2.37, m	2.30, m
9β	2.22, m	2.48, m	2.31, m	2.37, m	2.22, m
10	3.19 (d, 13.0)	2.28, m	3.37 (d, 13.1)	3.03 (d, 16.5)	3.06 (d, 16.5)
12α	1.50, m	1.43, m	1.45, m	1.46, m	1.44, m
12β	1.63, m	1.86, m	1.65, m	1.46, m	1.44, m
13α	2.25 (dd, 14.8, 6.3)	2.48, m	2.31, m	2.40, m	2.37, m
13β	2.59, m	2.95, m	2.58 (dd, 14.8, 6.4)	3.08, m	3.03, m
16	6.42 (d, 15.3)	6.37 (d, 15.1)	6.43 (d, 15.3)	6.31 (d, 15.3)	6.14 (d, 15.1)
17	6.38 (dd, 15.3, 9.7)	6.39 (dd, 15.1, 10.1)	6.40(dd, 15.3, 9.3)	6.34 (dd, 15.3, 10.3)	6.31 (dd, 15.1, 10.5)
18	5.93 (d, 9.7)	5.92 (d, 10.1)	5.90 (d, 9.3)	5.88 (d, 10.3)	5.94 (d, 10.5)
20	1.47, s	1.37, s	2.12, s	2.29, s	2.27, s
21	9.23, s	9.28, s	9.25, s	9.46, s	9.44, s
22	0.94, s	1.05, s	0.99, s	0.86, s	0.84, s
23	1.83, s	1.96, s	1.83, s	1.72, s	1.72, s
24	1.81, s	1.81, s	1.78, s	1.79, s	1.77, s
25	1.79, s	1.81, s	1.79, s	1.80, s	1.80, s

The assignments were aided by COSY, NOESY, HSQC, and HMBC NMR spectra. [a] Measured in CDCl$_3$; [b] Measured in methanol-d_4.

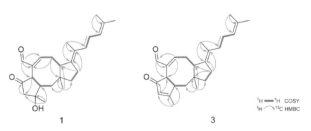

Figure 2. Key COSY and HMBC correlations of **1** and **3**.

Table 2. ^{13}C NMR data for **1–5** at 150 MHz (δ in ppm).

Position	1 [a]	2 [b]	3 [b]	4 [b]	5 [b]
1	41.4	41.5	45.3	34.9	34.9
2	49.6	50.6	49.5	49.8	49.9
3	76.7	76.6	179.9	180.2	180.2
4	55.0	53.9	129.4	130.1	130.1
5	216.6	216.2	209.4	209.2	209.4
6	48.8	48.8	49.7	48.0	48.0
7	142.1	141.8	140.6	138.7	138.6
8	158.4	160.7	156.7	159.7	160.0
9	34.1	29.3	34.0	28.8	29.9
10	47.8	56.2	47.2	41.9	40.7
11	43.6	44.2	44.0	45.1	45.3
12	44.5	34.9	43.4	40.4	40.4
13	27.2	26.9	26.8	31.4	33.1
14	146.8	143.3	146.6	143.2	142.7
15	124.9	126.5	125.0	127.1	126.6
16	130.6	131.1	130.6	130.0	129.6
17	123.6	122.7	123.2	122.9	123.9
18	125.9	126.0	126.0	126.0	126.1
19	134.9	133.7	133.5	133.6	133.8
20	25.8	24.4	15.7	17.2	17.1
21	194.0	195.1	193.2	195.1	195.2
22	21.3	18.0	19.8	24.6	24.7
23	13.6	14.1	13.8	14.2	15.5
24	18.4	16.9	16.9	16.9	16.9
25	26.1	24.8	24.8	24.8	24.8

[a] Measured in CDCl$_3$; [b] Measured in methanol-d_4.

The stereochemistry of **1** was determined by the analysis of proton-proton coupling constants and NOESY data. The strong NOESY correlations of H-6/H-10 and H-2/H$_3$-20/H$_3$-22 suggested that H-6 and H-10 were on the same face, and H-2, H$_3$-20 and H$_3$-22 were on opposite faces. Based on a comprehensive literature review, ophiobolin analogs have A/B-*cis* or A/B-*trans* isomers at C-2 and C-6 [18]. It has been reported that H-2 of the 6-epimer having H-6α is shielded by ca. 0.2-0.3 ppm in comparison with the A/B-cis isomer having H-6β [13]. On the basis of this analysis, the H-2 protons of **1** and **2** were observed at δ_H 2.19 and δ_H 2.57, respectively, indicating that **1** has an A/B-*trans* ring structure. The lack of a NOESY correlation of H-2/H-6 and a comparison of the ^{1}H NMR data of **1** with those of 6-*epi*-ophiobolin C (**6**) also supported the fact that the A/B ring junction is *trans* in **1** (Figure 3). The relative configuration of the side chain was confirmed by comprehensive NOESY and ^{1}H NMR analyses. The NOESY correlations of H-9α/H$_3$-23 and H-13β/H-16 indicated the relative configuration of $\Delta^{14,15}$ was *E* conformation (Figure 3a). The geometry of the $\Delta^{16,17}$ was confirmed as *E* by the large coupling constants of H-16 (d, *J* = 15.3 Hz) and H-17 (dd, *J* = 15.3, 9.7 Hz) and NOESY correlations of H-13β/H-16, H-17/H$_3$-23 and H-16/H-18. Moreover, a combination of literature review and comparison of the NMR spectral data and spectral properties of **1** with those of **6**, suggested that 14, 15-dehydro-6-*epi*-ophiobolin K (**1**) has the same absolute configuration of a 5-8-5 core structure in 6-*epi*-ophiobolin K [13,19,20].

Compound **2** had the same molecular formula C$_{25}$H$_{32}$O$_3$ as **1**. Its ^{1}H and ^{13}C NMR data were similar to those of **1**, differing only by slightly shifted proton and carbon signals. It has been reported that H-2 of the 6-*epi* isomer having H-6α (A/B-*trans*) is upfield-shifted in comparison with the A/B-*cis* ophiobolin [13] (Figure 3b). The H-2 proton (δ_H 2.57) of **2** is downfield-shifted than that (δ_H 2.19) of **1**, indicating that **2** has an A/B-*cis* ring structure. This study also revealed that the chemical shifts of the geminal proton H$_2$-4 are closer to each other when the A/B ring junction is *cis* than when it is *trans* (Figure 3b). The key NOE correlations of H-2/H-6, H-2/H$_3$-20 and H-2/H$_3$-22 suggested that **2** has an

A/B-*cis* ring structure and is a stereoisomer of **1** (Figure 3a). Based on these results, the structure of **2** was determined and named 14,15-dehydro-ophiobolin K.

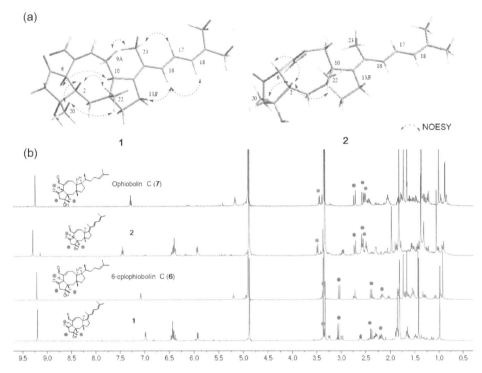

Figure 3. (**a**) Key NOESY correlations of **1** and **2**. (**b**) Comparison of chemical shifts of H_2-4 and H-6 in **1** (H-6α), **2** (H-6β), **6** (H-6α) and **7** (H-6β).

Compound **3** was obtained as an amorphous powder with the molecular formula of $C_{25}H_{32}O_2$ based on HRESIMS. The molecular formula of **3** has one less CH_2 and one less oxygen compared to that of **1**. The ^1H and ^{13}C NMR data of **3** were quite similar to those of **1**, displaying one additional singlet olefin signal (δ_H 6.01) and a sp^2 quaternary carbon at C-3 (δ_C 179.9), while lacking a methylene and sp3 quaternary carbon signal. The HMBC correlations from H_3-20 (δ_H 2.12) to C-2 (δ_C 49.5), C-3 (δ_C 179.9) and C-4 (δ_C 129.4) revealed that a double bond existed between C-3 and C-4 by the dehydroxylation of the tertiary alcohol at C-3 in **1** (Figure 2). NOESY correlations from H-6/H-10 and H-2/H_3-20/H_3-22, the lack of NOESY correlation of H-2/H-6 and, comparison of the NMR spectral data and spectral properties of **3** with those of 6-*epi*-ophiobolin N (**8**), suggested that **3** has the same ring system as the A/B-*trans* ophiobolin [13,19] (Figure 4). On the basis of detailed data analysis, the structure of **3** was elucidated and named 14,15-dehydro-6-*epi*-ophiobolin G.

Compound **4** was isolated as an amorphous powder with the molecular formula of $C_{25}H_{32}O_2$ as determined by HRESIMS. Its ^1H NMR data were similar to those of **3**, differing only by slightly shifted signals. In contrast to the data for **3**, the NOESY correlation of H-2/H-6 indicated that **4** has an A/B-*cis* ring structure (Figure 4). Thus, compound **4** was identified as a stereoisomer of **3** and named 14,15-dehydro-ophiobolin G.

Compound **5** was isolated as an amorphous powder with the same molecular formula $C_{25}H_{32}O_2$ as compound **4**, as determined by HRESIMS. The ^1H NMR data of **5** and **4** were nearly identical except for the H-16 proton which was slightly downfield-shifted than that of **4**. By comprehensive analysis of its 1D and 2D NMR data, the planar structure of **5** was elucidated to be the same as that of **4**, differing only in the orientation of H$_3$-23. The NOESY correlations of H-16/H-10 and H-23/H$_2$-13 suggested the relative configuration of $\Delta^{14,15}$ in **5** was *Z* conformation, which is different from that of compound **4** (Figure 4). Therefore, the structure of **5** was elucidated to be as shown in Figure 1, and named 14, 15-dehydro-(*Z*)-14-ophiobolin G.

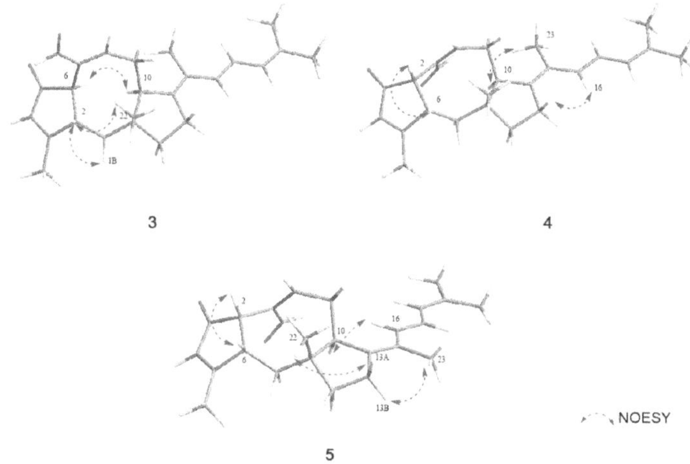

3   4

5

Figure 4. Key NOESY correlations of **3–5**.

The structures of the four known compounds were determined as 6-*epi*-ophiobolin C (**6**), ophiobolin C (**7**), 6-*epi*-ophiobolin N (**8**) and ophiobolin N (**9**) by comparing of their ^{1}H, ^{13}C NMR and MS data with those reported in literature (Supplementary Materials).

The cytotoxicity of all the isolated compounds (**1–9**) against cancer cell lines, such as HCT-15, NUGC-3, NCI-H23, ACHN, PC-3 and MDA-MB-231, was investigated using the sulforhodamine B (SRB) assay, with adriamycin as a positive control. The results showed that all compounds were strongly active against 6 cancer cell lines with GI$_{50}$ values in the range of 0.14 to 2.01 μM (Table 3). Compound **1** displayed the strongest cytotoxicity against the HCT-15, NUGC-3 and MDA-MB-231 cell lines with GI$_{50}$ values of 0.21, 0.19 and 0.14 μM, respectively. Based on the cytotoxicity results, the analogs with one double bond (**6–9**) in the side chain seemed to be slightly more active than those with three double bonds (**1–5**). **5** was least active against all cell lines, even with GI$_{50}$ values ranging from 1.53 to 2.01 μM, indicating that the geometry of C-14/C-15 might appear to have a slight influence on their activities. In addition, results for all the strongly active compounds indicated that the stereochemistry of C-6 and the hydroxyl group at C-3 might not noticeably affect the cytotoxicity.

Table 3. Growth Inhibition (GI$_{50}$, μM) Values of **1–9** against Human Tumor Cell Lines.

Cell Lines [a]	GI$_{50}$ (μM)									
	1	2	3	4	5	6	7	8	9	ADR [b]
HCT-15	0.21	0.44	0.96	1.24	1.67	0.24	0.21	0.30	0.22	0.13
NUGC-3	0.19	0.50	0.88	1.07	1.53	0.22	0.20	0.22	0.20	0.15
NCI-H23	0.18	0.61	1.40	1.50	1.84	0.24	0.16	0.22	0.22	0.15
ACHN	0.24	0.53	1.14	1.40	2.01	0.43	0.20	0.23	0.42	0.16
PC-3	0.24	0.47	1.00	1.38	1.60	0.27	0.36	0.20	0.20	0.14
MDA-MB-231	0.14	0.63	1.05	1.35	1.75	0.19	0.22	0.21	0.19	0.15

[a] HCT-15: Colon cancer, NUGC-3: Stomach cancer, NCI-H23: Lung cancer, ACHN: Renal cancer, PC-3: Prostate cancer, MDA-MB-231: Breast cancer; GI$_{50}$ values are the concentration corresponding to 50% growth inhibition.
[b] ADR: Adriamycin as standard.

3. Materials and Methods

3.1. General Experimental Procedures

The 1D (^{1}H and ^{13}C) and 2D (COSY, HSQC, HMBC, and NOESY) NMR spectra were obtained on a Bruker 600 MHz spectrometer. Specific optical rotations were obtained on a Rudolph Research Analytical (Autopol III) polarimeter. UV-visible spectra were acquired on a Shimadzu UV-1650PC spectrophotometer in 1 mm quartz cells. IR spectra were recorded on a JASCO FT/IR-4100 spectrophotometer. High-resolution ESIMS were recorded on a hybrid ion-trap time-of-flight mass spectrometer (Shimadzu LC/MS-IT-TOF). HPLC was performed using a PrimeLine Binary pump with RI-101(Shodex). RP-HPLC was performed using a semi-prep ODS column (YMC-Triart C18, 250 × 10 mm i.d, 5 μm) and an analytical ODS column (YMC-Triart C18, 250 × 4.6 mm i.d, 5 μm).

3.2. Fungal Material and Fermentation

The fungus 168ST-16.1 was isolated from the algae *Padina* sp., collected at a depth of 10 m in Son Tra peninsular, Da Nang, Vietnam (16°09′97.8″ N, 108°29′96.1″ E), in August 2016. The fungal strain was identified as *Aspergillus flocculosus* (GenBank accession number MG920345) by DNA amplification and ITS region sequencing and named *Aspergillus flocculosus* 168ST-16.1.

The isolated fungi were cultured on rice media at 28 °C for three weeks in 100 Erlenmeyer flasks (500 mL), each containing rice (20.0 g), yeast extract (20.0 mg), KH$_2$PO$_4$ (10 mg), and natural sea water (40 mL).

3.3. Isolation of Compounds 1–9

The whole fermentation media were extracted with EtOAc and evaporated *in vacuo* to give the crude extract (22 g), which was fractionated by flash column chromatography on ODS using a gradient of MeOH/ H$_2$O (1:4, 2:3, 3:2, 4:1 and 100% MeOH, each fraction 300× 3). The second fraction eluted with 100% MeOH was separated into ten subfractions (Fr. A-J) by column chromatography on ODS eluting with a step gradient of MeCN/H$_2$O (70:30 to 100:0, v/v). Fr. E (200 mg) was further purified by an analytical reversed-phase HPLC (YMC-Pack-ODS-A, 250 × 4.6 mm i.d, 5 μm, flow rate 2.5 mL/ min, isocratic elution with 55% MeCN in H$_2$O, RI detector) to yield 1 (7.5 mg, t_R = 18 min) and 2 (1.5 mg, t_R = 21 min). Compounds 3 (2.2 mg, t_R = 30 min), 4 (3.1 mg, t_R = 33 min), 5 (1.2 mg, t_R = 36 min), 6 (3.5 mg, t_R = 50 min) and 7 (3.2 mg, t_R = 53 min) were isolated from Fr. F (210 mg) by a semi-preparative reversed-phase HPLC (YMC-Pack-ODS-A, 250 × 10 mm i.d, 5 μm, flow rate 6.0 mL/ min, isocratic elution with 60% MeCN in H$_2$O, RI detector). Fr G (136 mg) was subjected to a semi-preparative reversed-phase HPLC (YMC-Pack-ODS-A, 250 × 10 mm i.d, 5 μm, flow rate 5.5 mL/ min, isocratic elution with 65% MeCN in H$_2$O, RI detector) to obtain 8 (3.6 mg, t_R = 40 min) and 9 (2.9 mg, t_R = 43 min).

14,15-dehydro-6-*epi*-ophiobolin K (1): amorphous powder; $[\alpha]_D^{23}$ +74.0(c 1.0, MeOH); IR ν_{max} 3442, 2931, 2852, 1736, 1683, 1640, 1454, 1379 cm^{-1}; UV(MeOH) λ_{max} (log ε) 286 (3.62), 236 (3.04) nm; HRESIMS *m/z* 405.2405 [M + Na]$^+$ (calcd for 405.2406, C$_{25}$H$_{34}$O$_3$Na); ^{1}H NMR (CDCl$_3$, 600 MHz) and ^{13}C NMR (CDCl$_3$, 150 MHz) see Table 1.

14,15-dehydro-ophiobolin K (2): amorphous powder; $[\alpha]_D^{23}$ +94.0(c 1.0, MeOH); IR ν_{max} 3451, 2967, 2897, 1734, 1688, 1448, 1377, 1233 cm^{-1}; UV(MeOH) λ_{max} (log ε) 289 (3.58), 238 (3.36) nm; HRESIMS *m/z* 405.2404 [M + Na]$^+$ (calcd for 405.2406, C$_{25}$H$_{34}$O$_3$Na); ^{1}H NMR (CD$_3$OD, 600 MHz) and ^{13}C NMR (CD$_3$OD, 150 MHz) see Table 1.

14,15-dehydro-6-*epi*-ophiobolin G (3): amorphous powder; $[\alpha]_D^{23}$ +87.0(c 1.0, MeOH); IR ν_{max} 2922, 2858, 1683, 1625, 1455, 1377 cm^{-1}; UV(MeOH) λ_{max} (log ε) 286 (3.53), 227 (3.24) nm; HRESIMS *m/z* 387.2301 [M + Na]$^+$ (calcd for 387.2300, C$_{25}$H$_{32}$O$_2$Na); ^{1}H NMR (CD$_3$OD, 600 MHz) and ^{13}C NMR (CD$_3$OD, 150 MHz) see Table 1.

14,15-dehydro-ophiobolin G (**4**): amorphous powder; $[\alpha]_D^{23}$ +85.0(c 1.0, MeOH); IR ν_{max} 2925, 2858, 1689, 1636, 1441, 1377 cm^{-1}; UV(MeOH) λ_{max} (log ε) 291 (3.59), 231 (3.37) nm; HRESIMS *m/z* 387.2299 [M + Na]$^+$ (calcd for 387.2300, $C_{25}H_{32}O_2Na$); ^{1}H NMR (CD$_3$OD, 600 MHz) and ^{13}C NMR (CD$_3$OD, 150 MHz) see Table 1.

14,15-dehydro-(Z)-14-ophiobolin G (**5**): amorphous powder; $[\alpha]_D^{23}$ +132.0(c 1.0, MeOH); IR ν_{max} 2922, 2855, 1692, 1632, 1437, 1377 cm^{-1}; UV(MeOH) λ_{max} (log ε) 289 (3.61), 231 (3.54) nm; HRESIMS *m/z* 387.2299 [M + Na]$^+$ (calcd for 387.2300, $C_{25}H_{32}O_2Na$); ^{1}H NMR (CD$_3$OD, 600 MHz) and ^{13}C NMR (CD$_3$OD, 150 MHz) see Table 1.

3.4. Cytotoxicity Test by SRB Assay

The human cancer cell lines, HCT-15 (colon), NUGC-3 (stomach), NCI-H23 (lung), ACHN (renal), PC-3 (prostate) and MDA-MB-231 (breast), were purchased from American Type Culture Collection (Manassas, VA, USA). They were then cultured in RPMI 1640 supplemented with 10% fetal bovine serum (FBS). Cell cultures were maintained at 37 °C under a humidified atmosphere of 5% CO_2. The growth inhibition assay against human cancer cell lines was performed in accordance with the sulforhodamine B (SRB) assay [22]. In brief, 8,000 cells/well were seeded onto a 96-well plate. On the following, the cells were treated with compounds **1–9**, vehicle control (0.1% DMSO) and positive control (adriamycin). After incubation for 48 h, the cultures were fixed with 50% trichloroactetic acid (50 μg/mL) and stained with 0.4% sulforhodamine B in 1% acetic acid. Unbound dye was removed by washing with 1% acetic acid, and protein-bound dye was extracted with 10 mM Tris base (pH 10.5) for optical density determination. Absorbance at 540 nm was determined using a VersaMax microplate reader from Molecular Devices (LLC, Sunnyvale, CA, USA). GI$_{50}$ values were calculated using GraphPad Prism 4.0 software from GraphPad Software, Inc. (San Diego, CA, USA).

4. Conclusions

Chemical investigation of the marine-derived fungus *Aspergillus flocculosus* 168ST-16.1 led to the isolation and identification of five new (**1–5**) and four known (**6–9**) ophiobolin derivatives. The five new ophiobolins possessed a fully unsaturated side chain, and **5** had a Z-conformation at C-14/C-15. To the best of our knowledge, the new compounds **1–5** are the first ophiobolins with three double bonds at the side chain. All compounds (**1–9**) exhibited potent growth inhibitory activities against the HCT-15, NUGC-3, NCI-H23, ACHN, PC-3 and MDA-MB-231 cancer cell lines. The cytotoxicities of the new ophiobolins **4** and **5** were slightly weaker or similar to those of the known compounds (**6–9**). These results suggest that dehydration at C-14 and C-15 might not significantly affect the cytotoxicity against cancer cell lines. This study is the first report to describe the effect of the side chain of ophiobolins by evaluating the anticancer activity of five new and four known compounds together.

Supplementary Materials: The followings are available online at http://www.mdpi.com/1660-3397/17/6/346/s1, Figures S1–S35: HRESI-MS data, ^{1}H NMR, ^{13}C NMR, COSY, HSQC, HMBC, NOESY and experimental spectra of **1–5**, Figures S36–S43: LRMS data, ^{1}H NMR, ^{13}C NMR and experimental spectra of **6–9**.

Author Contributions: H.J.S. was the principal investigator, who proposed ideas for the present work, managed and supervised the whole research work, prepared and corrected the manuscript, and contributed to the structure elucidation of the new and known compounds. B.-K.C. achieved all experiments for compounds **1–9**, including fermentation, isolation, and structure elucidation, and prepared the manuscript. P.T.H.T., H.-S.L., B.-W.C., N.T.D.N. and T.T.T.V. contributed to analyzing data. J.S.K. evaluated the cytotoxicity of **1–9**.

Funding: This research was supported in part by the Korea Institute of Ocean Science and Technology (Grant PE99752 to H.J.S.).

Acknowledgments: The authors express gratitude to Young Hye Kim, Korea Basic Science Institute, Ochang, Korea, for providing mass data.

Conflicts of Interest: The authors declare no conflict of interest.

References

1. Deshmukh, S.K.; Prakash, V.; Ranjan, N. Marine Fungi: A Source of Potential Anticancer Compounds. *Front. Microbiol.* **2018**, *8*, 2536. [CrossRef] [PubMed]
2. Blunt, J.W.; Carroll, A.R.; Copp, B.R.; Davis, R.A.; Keyzers, R.A.; Prinsep, M.R. Marine natural products. *Nat. Prod. Rep.* **2018**, *35*, 8–53. [CrossRef] [PubMed]
3. Debbab, A.; Aly, A.H.; Lin, W.H.; Proksch, P. Bioactive compounds from marine bacteria and fungi. *Microb. Biotechnol.* **2010**, *3*, 544–563. [CrossRef]
4. Saleem, M.; Ali, M.S.; Hussain, S.; Jabbar, A.; Ashraf, M.; Lee, Y.S. Marine natural products of fungal origin. *Nat. Prod. Rep.* **2007**, *24*, 1142–1152. [CrossRef] [PubMed]
5. Proksch, P.; Putz, A.; Ortlepp, S.; Kjer, J.; Bayer, M. Bioactive natural products from marine sponges and fungal endophytes. *Phytochem. Rev.* **2010**, *9*, 475–489. [CrossRef]
6. Bhadury, P.; Mohammad, B.T.; Wright, P.C. The current status of natural products from marine fungi and their potential as anti-infective agents. *J. Ind. Microbiol. Biotechnol.* **2006**, *33*, 325–337. [CrossRef] [PubMed]
7. Javed, F.; Qadir, M.I.; Janbaz, K.H.; Ali, M. Novel drugs from marine microorganisms. *Crit. Rev. Microbiol.* **2011**, *37*, 245–249. [CrossRef] [PubMed]
8. Molinski, T.F.; Dalisay, D.S.; Lievens, S.L.; Saludes, J.P. Drug development from marine natural products. *Nat. Rev. Drug. Discov.* **2009**, *8*, 69–85. [CrossRef]
9. Imhoff, J.F. Natural Products from Marine Fungi—Still an Underrepresented Resource. *Mar. Drugs.* **2016**, *14*, 19. [CrossRef]
10. Nicoletti, R.; Vinale, F. Bioactive Compounds from Marine-Derived *Aspergillus*, *Penicillium*, *Talaromyces* and *Trichoderma* Species. *Mar. Drugs.* **2018**, *16*, 408. [CrossRef]
11. Vadlapudi, V.; Borah, N.; Yellusani, K.R.; Gade, S.; Reddy, P.; Rajamanikyam, M.; Vempati, L.N.S.; Gubbala, S.P.; Chopra, P.; Upadhyayula, S.M.; et al. *Aspergillus* Secondary Metabolite Database, a resource to understand the Secondary metabolome of *Aspergillus* genus. *Sci. Rep.* **2017**, *7*, 7325. [CrossRef] [PubMed]
12. Trianto, A.; Widyaningsih, S.; Radjasa, O.K.; Pribadi, R. Symbiotic Fungus of Marine Sponge *Axinella* sp. Producing Antibacterial Agent. *Environ. Earth. Sci.* **2017**, *55*, 012005. [CrossRef]
13. Wei, H.; Itoh, T.; Kinoshita, M.; Nakai, Y.; Kurotaki, M.; Kobayashi, M. Cytotoxic sesterterpenes, 6-epi-ophiobolin G and 6-epi-ophiobolin N, from marine derived fungus *Emericella variecolor* GF10. *Tetrahedron* **2004**, *60*, 6015–6019. [CrossRef]
14. Au, T.K.; Chick, W.S.; Leung, P.C. The biology of ophiobolins. *Life Sci.* **2000**, *67*, 733–742. [CrossRef]
15. Bladt, T.T.; Frisvad, J.C.; Knudsen, P.B.; Larsen, T.O. Anticancer and antifungal compounds from Aspergillus, Penicillium and other filamentous fungi. *Molecules* **2013**, *18*, 11338–11376. [CrossRef] [PubMed]
16. Pak, C.L.; William, A.T.; Wang, J.H.; Carl, L.T. Role of Calmodulin Inhibition in the Mode of Action of Ophiobolin A. *Plant Physiol.* **1985**, *77*, 303–308.
17. Chai, H.; Yin, R.; Liu, Y.; Meng, H.; Zhou, X.; Zhou, G.; Bi, X.; Yang, X.; Zhu, T.; Zhu, W.; et al. Sesterterpene ophiobolin biosynthesis involving multiple gene clusters in *Aspergillus ustus*. *Sci. Rep.* **2016**, *6*, 27181. [CrossRef]
18. Tian, W.; Deng, Z.; Hong, K. The Biological Activities of Sesterterpenoid-Type Ophiobolins. *Mar. Drugs.* **2017**, *15*, 7.
19. Tsipouras, A.; Adefarati, A.A.; Tkacz, J.S.; Frazier, E.G.; Rohrer, S.P.; Birzin, E.; Rosegay, A.; Zink, D.L.; Goetz, M.A.; Singh, S.B.; et al. Ophiobolin M and analogues, noncompetitive inhibitors of ivermectin binding with nematocidal activity. *Bioorg. Med. Chem.* **1996**, *4*, 531–536. [CrossRef]
20. Bladt, T.T.; Dürr, C.; Knudsen, P.B.; Kildgaard, S.; Frisvad, J.C.; Gotfredsen, C.H.; Seiffert, M.; Larsen, T.O. Bio-Activity and Dereplication-Based Discovery of Ophiobolins and Other Fungal Secondary Metabolites Targeting Leukemia Cells. *Molecules* **2013**, *18*, 14629–14650. [CrossRef]
21. Nozoe, S.; Hirai, K.; Tsuda, K. The structure of zizanin-A and -B, C25-terpenoids isolated from helminthosporiumzizaniae. *Tetrahedron Lett.* **1966**, *7*, 2211–2216. [CrossRef]
22. Skehan, P.; Storeng, R.; Scudiero, D.; Monks, A.; McMahon, J.; Vistica, D.; Warren, J.T.; Bokesch, H.; Kenney, S.; Boyd, M.R. New colorimetric cytotoxicity assay for anticancer-drug screening. *J. Natl. Cancer. Inst.* **1990**, *82*, 1107–1112. [CrossRef] [PubMed]

Article

Isolation, Identification of Carotenoid-Producing *Rhodotorula* sp. from Marine Environment and Optimization for Carotenoid Production

Yanchen Zhao [1], Liyun Guo [2], Yu Xia [3], Xiyi Zhuang [1,*] and Weihua Chu [1,*]

[1] Department of Pharmaceutical Microbiology, School of Life Science and Technology, China Pharmaceutical University, Nanjing 210009, China; 17712936602@163.com
[2] Department of Microbiology, Nanjing Institute of Fisheries Science, Nanjing 210036, China; lyguo801@163.com
[3] Bureau of Ocean and Fisheries of Jiangsu Province, Nanjing 210003, China; xiayu6858@163.com
* Correspondence: zhuangtony@163.com (X.Z.); chuweihua@cpu.edu.cn (W.C.); Tel.: +86-025-8618-5398 (W.C.)

Received: 20 January 2019; Accepted: 4 March 2019; Published: 8 March 2019

Abstract: Carotenoids are natural pigments found in plants and microorganisms. These important nutrients play significant roles in animal health. In contrast to plant production, the advantages of microbial fermentation of carotenoids are the lower media costs, fast growth rate of microorganisms, and the ease of culture condition control. In this study, a colony of red pigment-producing yeast, *Rhodotorula* sp. RY1801, was isolated from the sediment of marine environment with the potential to produce carotenoids. Optimization of carotenoid production in *Rhodotorula* sp. RY1801 was also discussed. The optimum conditions found for carotenoid production were as follows: temperature, 28 °C; pH 5.0; carbon source, 10 g/L glucose, nitrogen source, 10 g/L yeast extract, maximum concentration of 987 µg/L of total carotenoids was obtained. The results of this study show that the isolated yeast strain *Rhodotorula* sp. RY1801 can potentially be used in future as a promising microorganism for the commercial production of carotenoids.

Keywords: carotenoids; optimization; red yeast; *Rhodotorula* sp.

---

1. Introduction

Carotenoids are pigments that exist in a wide variety of plants and microorganisms. They are characterized by yellow, orange, red or purple coloration [1]. Carotenoids have been proven to play important roles in animal health as precursors of vitamin A, scavengers of active oxygen, and enhancers of in vitro antibody production. Therefore, they are widely applied in animal feed additives as nutrient supplements, food, pharmaceutical, and cosmetic industries as dyes/colorants and functional ingredients [2]. Carotenoids are in high demand throughout the world, so a suitable method for an industrial production of carotenoids producing is needed. Most of the carotenoids are extracted from plants like annatto, tomato, grapes, carrot, paprika, etc. Carotenoids can also be produced from microorganisms [3].

Carotenoids can be produced by numerous microorganisms. Filamentous fungi, yeasts, bacteria and algae, such as *Streptomyces chrestomyceticus*, *Blakeslea trispora*, *Phycomyces blakesleeanus*, *Flavobacterium* sp., *Phaffia* sp., and *Rhodotorula* sp., Actinomycetes have been described as carotenoid-producing microorganisms [4,5]. The production of carotenoids from microorganisms have advantages over plants, such as higher yields, less batch-to-batch variations, easily manipulated, and no seasonal or geographic variations [6,7]. Carotenoid producing microorganisms, such as bacteria and archaea, algae and fungi; are abundant in the natural environment. Microalgae are currently the main sources of industrial carotenoid production [8], but other microorganisms could be valid alternatives [9]. With the rising demand of carotenoids, there has been renewed interest in identifying novel carotenoid-producing microorganisms. Yeast is the most suitable candidate for carotenoid production because of its fast growth rate, and the ease of cultivation. Yeast has the potential to produce large amounts of carotenoids such as lycopene, β-carotene, astaxanthin, torulene and torularhodin, etc. Carotenoid-producing yeasts are mainly represented by the genera *Rhodotorula* sp., *Rhodosporidium* sp., *Sporobolomyces* sp., *Xanthophylomyces* sp. [10,11]. Microorganisms that inhabit marine environments have been considered useful natural sources for new biomolecules production. Marine microorganisms possess unique metabolic and physiological features. They have evolved protective mechanisms compared to terrestrial microorganisms, which include the accumulation of bioactive compounds. It is also considered that the production of these bioactive compounds may be relatively easy by marine microorganisms [12,13]. The aim of the current study was carried out to isolate, identify carotenoid-producing strains from the marine environment, and optimize the nutritional and environmental parameters for their carotenoid production.

2. Results and Discussion

2.1. The Isolation and Identification of Carotenoids Producing Yeasts

A total of six morphologically distinct yeasts with red pigment were isolated from marine sediment samples as pure cultures (designated as RY1801–RY1806). Among the isolates, only the strain RY1801 had rapid growth and high pigment producing abilities, which was subsequently used for further study. The isolate RY1801 developed mucous, smooth surface and red-colored colonies on YPD agar plate (Figure 1A), and the growth was frequently observed in the microscopic examination (Figure 1B). Cells of the isolated strain RY1801 had an oval shape, the RY1801 cells size was 4–6.5 μm × 2–3.5 μm and had a colony diameter of 1.5 mm after 24 h cultivation (Figure 1B). The liquid medium changed to red after 24 h cultivation (Figure 1C). It has assimilated sugars such as glucose, galactose, sucrose, maltose, melezitose, and raffinose. The nitrate assimilation was positive. Further biochemical tests were carried out and are listed in Table 1. These biochemical results were not sufficient for classification to the genus, so the Internal Transcribed Spacers (ITS) of the ribosomal DNA sequence was amplified.

Figure 1. (**A**), Pure culture of the potential marine yeast strain *Rhodotorula* sp. RY1801 on YPD agar. (**B**), Micro-morphology of RY1801 observed under 40× with methylene blue staining. (**C**), Liquid culture of RY1801.

The ITS sequence obtained from strain *Rhodotorula* sp. RY1801 (GeneBank MH760806) was compared with the sequences in the GeneBank database and revealed that the strain RY1801 had 99% homology to *Rhodotorula babjevae*. The nucleotide sequence of the ITS region from RY1801 strain was identical to two other *R. babjevae* sequences included in the phylogenetic tree (Figure 2). Based on morphological, physiological characteristics, and ITS sequence, the isolated red yeast RY1801was tentatively named R. babjevae RY1801 and deposited at the China General Microbiological Culture Collection Center (CGMCC, Beijing) as CGMCC No. 15980.

Table 1. Morphological, physiological and biochemical characteristics of isolated yeast strain RY1801.

Assimilation Reactions	*Rhodotorula* sp. RY1801	Assimilation Reactions	*Rhodotorula* sp. RY1801	Assimilation Reactions	*Rhodotorula* sp. RY1801
Glucose	+	Ethanol	-	2-keto-D-gluconate	-
Galactose	+	Glycerol	+	Xylitol	-
Sucrose	+	Erythritol	-	50% glucose	-
Maltose	+	Ribitol	+	10% NaCl/5% Glucose	-
Cellobiose	-	Galactitol	+	Starch formation	-
Trehalose	+	D-Mannitol	+	Urease	+
Lactose	-	D-Glucitol	-	Gelatin liquefaction	-
Melibiose	-	α-Methyl D-glucose	+	Growth at 19 °C	+
Raffinose	+	Salicin	-	Growth at 25 °C	+
Melezitose	+	D-Gluconate	+	Growth at 37 °C	+
Inulin	-	DL-Lactate	+	Growth at 40 °C	-
Soluble starch	+	Succinate	+	Pellicle	-
D-Xylose	+	Citrate	+	Sedimentation	+
L-Arabinose	+	Inositol	-	True hyphae	-
D–Glucosamine	-	Hexadecane	+	Acid production	-
N-acetyl-D-glucosamine	-	Nitrate	+		
Methanol		Vitamin-free			

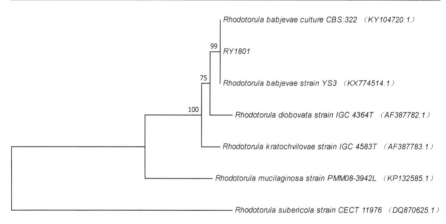

Figure 2. Phylogenetic tree of RY1801 obtained by neighbor-joining analysis of ITS region of rDNA.

Several yeast species can synthesize carotenoids, in particular, the genera *Xanthophyllomyces*, *Rhodotorula*, *Sporobolomyces*, and *Phaffia* have been used to produce carotenoids [14,15]. The production of carotenoid pigments in numerous natural isolates of the genera *Rhodotorula* has been studied by others, such as *Rhodotorula glutinis*, *Rhodotorula minuta*, *Rhodotorula mucilaginosa*, *Rhodotorula acheniorum* and *Rhodotorula graminis* [16]. El-Banna et al. isolated 46 yeast isolates from natural environments, all the strains belonged to *Rhodotorula glutinis* [17]. Muthezhilan et al. isolated a marine yeast *Rhodotorula* Sp. (Amby109) which can produce carotenoid pigments [18]. In this study, we isolated and identified the marine yeast which can produce red pigment, based on morphological, physiological characteristics and ITS sequence, and our results showed that the isolated red yeast RY1801 belonged to *Rhodotorula* sp.

The carotenoid pigments extracted from *Rhodotorula* sp. RY1801 have shown no inhibitory activity against all the detected strains. Muthezhilan et al. results showed that the pigment derived from marine yeast *Rhodotorula* Sp. (Amby109) have strong antimicrobial activity [18].

2.2. Effects of Various Parameters on Biomass Growth and Carotenoids Production

2.2.1. The Incubation Temperature

Incubation temperature ranging from 20 to 37 °C were checked for the biomass and carotenoids production in *Rhodotorula* sp. RY1801. As shown in Figure 3, the optimal temperature for biomass and carotenoids production was 28 °C, although reduced biomass and carotenoid production was seen and tested at other temperatures. Our results were similar to the results provided by others. Other studies revealed the optimal temperature for maximum *Rhodotorula glutinis* growth and carotenoids production was 29 and 30 °C, respectively [19] in monoculture and 30 °C in co-culture with lactic acid bacteria [20]. The temperature also has an effect on the regulation of enzymes involved in carotenoids production [21].

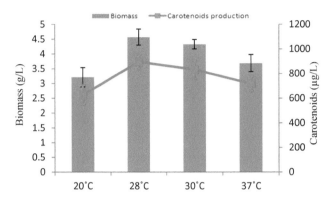

Figure 3. Effect of temperature on biomass and carotenoids production by *Rhodotorula* sp. RY1801.

2.2.2. Culture Medium pH

The influence of culture medium pH on biomass growth and carotenoids production in *Rhodotorula* sp. RY1801 was evaluated in YPD medium at 28 °C. As seen in Figure 4, the optimal initial pH under our culture conditions was pH 5 and similar biomass and carotenoids concentrations were seen at pH 6.0 and 7.0. Our results were similar to those by other workers. A study by Latha et al. indicated that the *R. glutinis* biomass increased as the initial culture pH increased from 5.5 to 7.5, although optimal carotenoids production was pH 5.5 [22]. A similar optimal pH 5.5 was observed for β-Carotene production in a related species, *Rhodotorula acheniorum* [23].

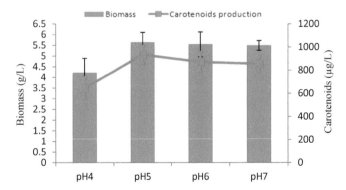

Figure 4. Effect of initial pHs on production of biomass and carotenoids by *Rhodotorula* sp. RY1801.

2.2.3. Carbon Sources

Carbon has been considered an important source for the energy supply and growth of microorganisms and is widely studied in the context of microbial fermentations. We investigated the influence of several carbon sources on biomass, and carotenoid production under culture conditions with initial pH 5.0 at 28 °C. Among the several carbon sources tested, glucose proved to be the most suitable carbon source for carotenoid production, with 962 µg/L of carotenoid (Figure 5). This may be due to the fact that glucose can easily be assimilated in the metabolic pathway for biosynthesis of carotenoids. The type of carbon source has a significant influence on carotenoids production, and their effects may differ depending on the yeast strains [24].

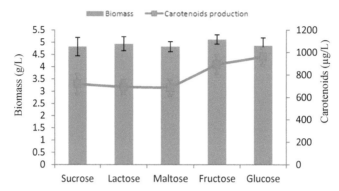

Figure 5. Effect of different carbon sources on production of biomass and carotenoids by *Rhodotorula* sp. RY1801.

2.2.4. Inorganic Nitrogen and Organic Nitrogen Sources

The influence of different nitrogen sources on biomass and carotenoids production was investigated with culture media containing 2% (*w/v*) glucose, initial pH 5.0, at 28 °C. Among the tested nitrogen sources, yeast extract was proved to be the most suitable nitrogen source for carotenoids production, with 987 μg/L of carotenoid (Figure 6). The influence of nitrogen sources on carotenoids production in *Rhodotorula* sp., also depend on the different strains.

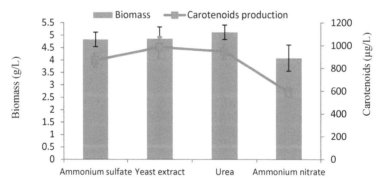

Figure 6. Effect of nitrogen sources on production of biomass and carotenoids by *Rhodotorula* sp. RY1801.

Optimization of cultural conditions is necessary in microbial fermentations for carotenoids to fully exploit the potential of selected microbial strain. The fermentation conditions for the production of carotenoids by the new isolated *Rhodotorula* sp. strain was optimized in shake flasks. With different culture conditions, the amount of biomass varied widely from 3.21 to 5.63 g/L and the total carotenoids content varied from 589 to 987 μg/L. The biomass yield in our study is lower than others, which could be attributed to the short time of culture (3 days). El-Banna et al. found that *Rhodotorula glutinis* strain NO. 0 produced 7 g/L dry biomass and 266 μg/g cellular carotenoids, 1.6 μg/L volumetric carotenoids after growing at 30 °C for 4 days [16]. Hamidid et al. have also reported that the production biomass was ranged from 0.04 to 0.84 g/L and the total carotenoid from 0.15 to 10.78 mg/L when optimizing culture conditions for *Halorubrum* sp. [25].

3. Materials and Methods

3.1. Sample Collection and Yeast Isolation

Different sediment samples were collected from the exposed intertidal zone along the South Yellow Sea in Dongtai City, Jiangsu Province, China. Each sediment sample (approximately 100 g) was placed in a sterile plastic bag with an ice bag and transported to the laboratory within 10 h and then processed immediately to isolate yeast. Ten grams of each sample (wet mass) were homogenized in 90 mL sterile 0.9% saline solution then individual yeast colonies were obtained by serial dilution and plating on yeast extract–peptone–dextrose (YPD) agar plates. All the plates were incubated at 28 °C for 24–48 h to determine the morphology of the colony.

3.2. Identification of the Red Yeast RY1801 Strain

The pure culture of strain RY1801 was used to investigate its physiological and morphological characteristics according to the methods described by Kurtzman et al. [26]. Genomic DNA of RY1801 was extracted using QIAamp DNA Mini Kit (QIAGEN) following the manufacturer's instructions. DNA amounts and purity contained in each extract were evaluated by measuring the absorbances at 230, 260 and 280 nm (Nanodrop 2000, Thermo Scientific, Waltham, MA, USA) and calculating the ratio A260/A280 and A260/A230. DNAs were stored at -20 °C prior to use for amplification studies [27]. The ITS region was sequenced using the forward primer ITS1 (5′-TCCGTAGGTGAACCTGCGG-3′) and the reverse primer ITS4 (5′-TCCTCCGCTTATTGATATGC-3′) [28]. The PCR conditions were as follows: 94 °C for 10 min, followed by 30 cycles, 92 °C for 1 min, 52 °C for 1 min, 72 °C for 1 min, and final synthesis at 72 °C for 5 min. The PCR products were separated by agarose gel electrophoresis and purified for sequencing. The sequences obtained were compared to rDNA sequences from the GeneBank (http://www.ncbi.nlm.nih.gov/BLAST/). ITS fragments obtained from GeneBank database were aligned with ClustalW (http://www.ebi.ac.uk/Tools/clustalw2/index.html) and the phylogenetic tree was computed with Jalview 2.4.0.b2 using the neighbor-joining method.

3.3. Determination of Biomass and Total Carotenoids

The cells were harvested by centrifugation 8000 rpm and 4 °C for 10 min, later washed with distilled water and centrifuged. The biomass of RY1801 was quantified through drying at 60 °C until a constant mass was obtained.

The carotenoids were extracted using techniques as described by Lopes et al. [29] with slight modification. The 0.1 g dry weight biomass was mixed with 2 mL DMSO and 5 mL acetone in a 10 mL tube. The mixture was subjected to 5 ultrasonic cycles at 40 kHz (Ningbo Scientz Biotechnology Co., Ltd., Ningbo, China) for 10 min, with 5 ml acetone added. The tube was vortexed vigorously and kept standing for 10 min. Centrifugation was performed (5000 × g for 10 min) to remove the biomass from the extracted carotenoids. The biomass was then resuspended in DMSO and acetone for additional extractions. The carotenoids-containing supernatant was pooled and analyzed by spectrophotometry. Initial spectrophotometry scan between 300 and 600 nm revealed the maximum absorption to occur at 490 nm. Carotenoids concentration was determined using the following equation [30–32].

Total carotenoids (μg/g of yeast) $= A_{max} \times D \times V/(E \times W)$

A_{max}: the absorbencies of total extract carotenoid at 490 nm

D: sample dilution ratio

V: volume of extraction solvent (mL)

E: extinction coefficient of total carotenoid (0.16)

W: dry weight of yeast (g)

3.4. Antimicrobial Activity of Carotenoid Pigments

The carotenoid pigments were extracted and dissolved in methanol. The antimicrobial activity was obtained using an agar well diffusion method [33]. *Escherichia coli* ATCC 29522, *Staphylococcus aureus* ATCC 25923 and *Pseudomonas aeruginosa* ATCC27853 were used as indicator strains. After incubation at 37 °C for 24 h, the antimicrobial activity of carotenoid pigments was determined by measuring the diameter of the zone of inhibition.

3.5. Optimization of Carotenoid Production in Shake-Flasks Experiments

In order to determine the initial pH values, incubation temperature, carbon source and nitrogen sources on carotenoids production and biomass growth, the experiment was conducted using a series of 250-mL flasks. Each flask contained 100 ml YPD media with 5% inoculum of *Rhodotorula* sp. RY1801. The initial media pH values, incubated temperatures, different carbon sources, and nitrogen sources were adjusted according to the experimental design. The flasks were shaken at 120 rpm for 72 h. The yeast biomass was harvested using refrigerated centrifugation (8000 rpm, 10 min). After washing the cellular pellet with distilled water twice, the biomass was used for further carotenoids extraction and carotenoids production analysis.

4. Statistical Analysis

All data were analyzed using One-way Analysis of Variance (ANOVA), and multiple comparison tests (Duncan's and Tukey's-tests) were performed using SPSS Statistic 2.0 software. Data were presented as Mean ± Standard. $p < 0.05$ was considered statistically significant.

5. Conclusions

Red yeast strain *Rhodotorula* sp. RY1801 was isolated from the exposed intertidal zone along the South Yellow Sea in Dongtai City, Jiangsu Province, China. The optimum conditions found for carotenoids production for *Rhodotorula* sp. RY1801 were as follows: temperature, 28 °C; pH 5.0; carbon source, 10 g/L glucose; and nitrogen source, 10 g/L yeast extract, maximum concentration of 987 µg/L of total carotenoids was obtained. The results of this study showed that the isolated yeast strain *Rhodotorula* sp. RY1801 potentially can be used in the future as promising microorganism for the commercial production of carotenoids.

Author Contributions: Planned the experiments: W.C. Performed the experiments: Y.Z., L.G. and X.Z. Analyzed the data: Y.Z. Contributed reagents or other essential material: W.C. and Y.X. Wrote the paper: X.Z. Read and approved the final manuscript: Y.Z., L.G., Y.X., X.Z. and W.C.

Funding: This research was funded by the Marine and Fishery Science and Technology Innovation Project of Jiangsu Ocean and Fishery Bureau, China (grant number Y2017-40) and the Priority Academic Program Development (PAPD) of Jiangsu Higher Education Institutions.

Acknowledgments: The authors are grateful to Tchoudjin Djeukwe Sybille and Robert JC McLean for English editing of the manuscript.

Conflicts of Interest: The authors declare no conflict of interest.

References

1. Botella-Pavía, P.; Rodríguez-Concepción, M. Carotenoid biotechnology in plants for nutritionally improved foods. *Physiol. Plant.* **2006**, *126*, 369–381. [CrossRef]
2. Jomova, K.; Valko, M. Health protective effects of carotenoids and their interactions with other biological antioxidants. *Eur. J. Med. Chem.* **2013**, *70*, 102–110. [CrossRef] [PubMed]
3. Sankari, M.; Rao, P.R.; Hemachandran, H.; Pullela, P.K.; Tayubi, I.A.; Subramanian, B.; Gothandam, K.M.; Singh, P.; Ramamoorthy, S. Prospects and progress in the production of valuable carotenoids: Insights from metabolic engineering, synthetic biology, and computational approaches. *J. Biotechnol.* **2018**, *266*, 89–101. [CrossRef] [PubMed]

4. Mussagy, C.U.; Winterburn, J.; Santos-Ebinuma, V.C.; Pereira, J.F.B. Production and extraction of carotenoids produced by microorganisms. *Appl. Microbiol. Biotechnol.* **2019**, *103*, 1095–1114. [CrossRef] [PubMed]
5. Lee, P.; Schmidt-Dannert, C. Metabolic engineering towards biotechnological production of carotenoids in microorganisms. *Appl. Microbiol. Biotechnol.* **2002**, *60*, 1–11. [CrossRef] [PubMed]
6. Mortensen, A. Carotenoids and other pigments as natural colorants. *Pure Appl. Chem.* **2006**, *78*, 1477–1491. [CrossRef]
7. Mapari, S.A.S.; Thrane, U.; Meyer, A.S. Fungal polyketide azaphilone pigments as future natural food colorants? *Trends Biotechnol.* **2010**, *28*, 300–307. [CrossRef]
8. Guedes, A.C.; Amaro, H.M.; Malcata, F.X. Microalgae as sources of carotenoids. *Mar. Drugs* **2011**, *9*, 625–644. [CrossRef]
9. Corinaldesi, C.; Barone, G.; Marcellini, F.; Dell'Anno, A.; Danovaro, R. Marine microbial-derived molecules and their potential use in cosmeceutical and cosmetic products. *Mar. Drugs* **2017**, *15*, 118. [CrossRef]
10. Mannazzu, I.; Landolfo, S.; da Silva, T.L.; Buzzini, P. Red yeasts and carotenoid production: Outlining a future for non-conventional yeasts of biotechnological interest. *World J. Microbiol. Biotechnol.* **2015**, *31*, 1665–1673. [CrossRef]
11. Kot, A.M.; Błażejak, S.; Gientka, I.; Kieliszek, M.; Bryś, J. Torulene and torularhodin: "New" fungal carotenoids for industry? *Microb. Cell Fact.* **2018**, *17*, 49. [CrossRef] [PubMed]
12. Torregrosa-Crespo, J.; Montero, Z.; Fuentes, J.L.; Reig García-Galbis, M.; Garbayo, I.; Vílchez, C.; Martínez-Espinosa, R.M. Exploring the valuable carotenoids for the large-scale production by marine microorganisms. *Mar. Drugs* **2018**, *16*, 203. [CrossRef] [PubMed]
13. Galasso, C.; Corinaldesi, C.; Sansone, C. Carotenoids from marine organisms: Biological functions and industrial applications. *Antioxidants* **2017**, *6*, 96. [CrossRef] [PubMed]
14. Ambati, R.R.; Phang, S.M.; Ravi, S.; Aswathanarayana, R.G. Astaxanthin: Sources, extraction, stability, biological activities and its commercial applications—A review. *Mar. Drugs* **2014**, *12*, 128–152. [CrossRef] [PubMed]
15. Mata-Gómez, L.C.; Montañez, J.C.; Méndez-Zavala, A.; Aguilar, C.N. Biotechnological production of carotenoids by yeasts: An overview. *Microb. Cell Fact.* **2014**, *13*, 12. [CrossRef] [PubMed]
16. Frengova, G.I.; Beshkova, D.M. Carotenoids from *Rhodotorula* and *PhaYa*: Yeasts of biotechnological importance. *J. Ind. Microbiol. Biotechnol.* **2009**, *36*, 163. [CrossRef]
17. El-Banna, A.; Abd El-Razek, A.; El-Mahdy, A. Isolation, identification and screening of carotenoid-producing strains of *Rhodotorula glutinis*. *Food Nutr. Sci.* **2012**, *3*, 627–633. [CrossRef]
18. Muthezhilan, R.; Ragul, R.; Pushpam, R.L.; Narayanan, R.L.; Hussain, A.J. Isolation, optimization and extraction of microbial pigments from marine yeast Rhodotorula Sp (Amby109) as food colourants. *Biosci. Biotechnol. Res. Asia* **2014**, *11*, 271–278. [CrossRef]
19. Malisorn, C.; Suntornsuk, W. Optimization of β-carotene production by *Rhodotorula glutinis* DM28 in fermented radish brine. *Bioresour. Technol.* **2008**, *99*, 2281–2287. [CrossRef]
20. Frengova, G.I.; Simova, E.D.; Beshkova, D.M. Effect of temperature changes on the production of yeast pigments co-cultivated with lacto-acid bacteria in whey ultrafiltrate. *Biotechnol. Lett.* **1995**, *17*, 1001–1006. [CrossRef]
21. Hayman, E.P.; Yokoyama, H.; Chichester, C.O.; Simpson, K.L. Carotenoid biosynthesis in *Rhodotorula glutinis*. *J. Bacteriol.* **1974**, *120*, 1339–1343. [PubMed]
22. Latha, B.V.; Jeevaratnam, K.; Murali, H.S.; Manja, K.S. Influence of growth factors on carotenoid pigmentation of *Rhodotorula glutinis* DFR-PDY from natural source. *Indian J. Biotechnol.* **2005**, *4*, 353–357.
23. Nasrabadi, M.R.N.; Razavi, S.H. Optimization of β-carotene production by a mutant of the lactose-positive yeast *Rhodotorula acheniorum* from whey ultrafiltrate. *Food Sci. Biotechnol.* **2011**, *20*, 445–454. [CrossRef]
24. Kot, A.M.; Błażejak, S.; Kurcz, A.; Gientka, I.; Kieliszek, M. *Rhodotorula glutinis*—Potential source of lipids, carotenoids, and enzymes for use in industries. *Appl. Microbiol. Biotechnol.* **2016**, *100*, 6103–6117. [CrossRef] [PubMed]
25. Hamidi, M.; Abdin, M.Z.; Nazemyieh, H.; Hejazi, M.A.; Hejazi, M.S. Optimization of total carotenoid production by *Halorubrum* Sp. TBZ126 using response surface methodology. *J. Microb. Biochem. Technol.* **2014**, *6*, 286–294. [CrossRef]
26. Kurtzman, C.P.; Fell, J.W.; Boekhout, T.; Robert, V. *The Yeasts, a Taxonomic Study*, 5th ed.; Elsevier: Burlington, MA, USA, 2011; pp. 87–110, ISBN 978-7-5496-2109-5.

27. Knebelsberger, T.; Stoger, I. DNA extraction, preservation, and amplification. *Methods Mol. Biol.* **2012**, *858*, 311–338. [CrossRef] [PubMed]
28. White, T.J.; Bruns, T.; Lee, S. Amplification and direct sequencing of fungal ribosomal RNA genes for phylogenetics. *PCR Protoc. Guide Methods Appl.* **1990**, *18*, 315–322.
29. Lopes, N.A.; Remedi, R.D.; dos Santos Sá, C.; André Veiga Burkert, C.; de Medeiros Burkert, J.F. Different cell disruption methods for obtaining carotenoids by *Sporodiobolus pararoseus* and *Rhodothorula mucilaginosa*. *Food Sci. Biotechnol.* **2017**, *26*, 759–766. [CrossRef]
30. Gu, Z.; Chen, D.; Han, Y.; Chen, Z.; Gu, F. Optimization of carotenoids extraction from *Rhodobacter sphaeroides*. *LWT-Food Sci. Technol.* **2008**, *41*, 1082–1088. [CrossRef]
31. Chen, D.; Han, Y.; Gu, Z. Application of statistical methodology to the opti- mization of fermentative medium for carotenoids production by *Rhodobacter sphaeroides*. *Process Biochem.* **2006**, *41*, 1773–1778. [CrossRef]
32. Cheng, Y.T.; Yang, C.F. Using strain *Rhodotorula mucilaginosa* to produce carotenoids using food wastes. *J. Taiwan Inst. Chem. Eng.* **2016**, *61*, 270–275. [CrossRef]
33. Nalawade, T.M.; Bhat, K.G.; Sogi, S. Antimicrobial activity of endodontic medicaments and vehicles using agar well diffusion method on facultative and obligate anaerobes. *Int. J. Clin. Pediatr. Dent.* **2016**, *9*, 335–341. [CrossRef] [PubMed]

Article

Metabolomic Insights into Marine Phytoplankton Diversity

Rémy Marcellin-Gros [1,2], Gwenaël Piganeau [2,*] and Didier Stien [1,*]

[1] Sorbonne Université, CNRS, Laboratoire de Biodiversité et Biotechnologie Microbiennes, LBBM, Observatoire Océanologique, 66650 Banyuls-sur-Mer, France; marcellin-gros@obs-banyuls.fr

[2] Sorbonne Université, CNRS, Biologie Intégrative des Organismes Marins, BIOM, Observatoire Océanologique, 66650 Banyuls-sur-Mer, France

* Correspondence: gwenael.piganeau@obs-banyuls.fr (G.P.); didier.stien@cnrs.fr (D.S.);
Tel.: +33-468887343 (G.P.); +33-430192476 (D.S.)

Received: 30 November 2019; Accepted: 22 January 2020; Published: 25 January 2020

Abstract: The democratization of sequencing technologies fostered a leap in our knowledge of the diversity of marine phytoplanktonic microalgae, revealing many previously unknown species and lineages. The evolutionary history of the diversification of microalgae can be inferred from the analysis of their genome sequences. However, the link between the DNA sequence and the associated phenotype is notoriously difficult to assess, all the more so for marine phytoplanktonic microalgae for which the lab culture and, thus, biological experimentation is very tedious. Here, we explore the potential of a high-throughput untargeted metabolomic approach to explore the phenotypic–genotypic gap in 12 marine microalgae encompassing 1.2 billion years of evolution. We identified species- and lineage-specific metabolites. We also provide evidence of a very good correlation between the molecular divergence, inferred from the DNA sequences, and the metabolomic divergence, inferred from the complete metabolomic profiles. These results provide novel insights into the potential of chemotaxonomy in marine phytoplankton and support the hypothesis of a metabolomic clock, suggesting that DNA and metabolomic profiles co-evolve.

Keywords: chemotaxonomy; phylogeny; mamiellales; galactolipids; betaine lipids; xanthophylls

1. Introduction

Phytoplanktonic eukaryotes are phylogenetically highly diverse, as they have many representatives in most super-groups of the eukaryotic tree of life [1,2]. The Archaeplastida super-group, or green lineage [3], includes all of the species that have descended from a primary endosymbiosis event, when an ancestral eukaryotic cell engulfed a photosynthetic prokaryote, that eventually evolved into an organelle, the chloroplast [4]. Our knowledge on the diversity of phytoplanktonic green microalgae has greatly increased with the democratization of DNA sequencing and genomics, but it is likely to stay behind that of their terrestrial relatives, including land plants, whose estimated species number exceeds 400,000 [5]. This is not surprising since unicellular organisms are generally less studied than multicellular organisms, and because bona fide species identification relies on tedious sampling and isolation steps which have yet to be performed for many marine microalgae. Recently, DNA sequencing of numerous environmental marine water sample extracts collected worldwide during the Ocean Sampling Day initiative provided evidence that many of the sequences detected belong to species from the green lineage that have no representative strains in culture [6]. This study also demonstrated the very broad geographic distribution of species from the class Mamiellophyceae, that dominates the picoeukaryotic fraction (cell diameter <3 μm) [7] in many coastal areas and thus plays a key ecological role in marine food webs. These picoalgae exemplify the ecological success of miniaturized eukaryotic cells [8] displaying a simple cellular organization (one mitochondrion and one chloroplast) and high surface–volume ratio, which is likely to confer advantages in nutrient-poor environments [9]. The Mamiellophyceae lineage is ancient, probably over 350 million years old [10], and currently comprises 22 described species [11]. Species from the three genera *Bathycoccus*, *Micromonas*, and *Ostreococcus* are particularly prevalent in the marine environment [6].

Historically, analysis of pigment composition has been used to assist the classification of microalgae. Indeed, *Bathycoccus*, *Micromonas*, and *Ostreococcus* species were found to contain characteristic pigments of many marine green microalgae, prasinoxanthin [12], but also specific pigments such as uriolide, micromonal, and dihydrolutein [13], which were not detected in other species outside the Mamiellales.

The advent of metabolomics now enables the metabolic signatures of microalgae to be explored at an unprecedented level of resolution [14]. These novel approaches have led to the discovery of new metabolites, fostered by the search for natural bioactive compounds with applications in agronomic, medical, or biofuel research. Indeed, polyunsaturated fatty acids (PUFA) are lipids with high nutritional value [15] and may have applications in the prevention of several pathologies, such as cancers or cardiovascular diseases [16,17]. Furthermore, the anti-inflammatory and antiviral properties of polar lipids have also been highlighted recently [18,19].

Here, we explored the potential of an untargeted metabolomic approach including pigments, lipids, and other uncharacterized metabolites to investigate chemotaxonomic markers in 12 marine microalgal strains from 11 species, including 9 microalgae from the green lineage; the Mamiellales *Ostreococcus tauri* [20], *O. mediterraneus* [21], *Bathycoccus prasinos* [22], *Micromonas commoda* [23], and *Mantoniella* sp., the Chlorellales *Picochlorum costavermella* [24], and strains from basal groups *Nephroselmis* sp. and *Pyramimonas* sp. To broaden the phylogenetic diversity of the dataset, two additional marine microalgae outside the green lineage were included: *Phaeodactylum tricornutum* (Stramenopile lineage) and *Pavlova lutheri* (Haptophyta lineage). We assess whether total metabolomic profiling enables us to delineate well-defined species, and we characterize the 10 major compounds of each species. Last but not least, we assess whether distances between metabolomic profiles, integrating both the composition and the frequency of each compound, reflect phylogenetic distances between species. This leads us to discuss the hypothesis of a metabolomic corollary of the molecular clock, a central tenet of molecular evolution.

2. Results

2.1. An Untargeted Holistic Analysis of Metabolomic Profiles

To investigate metabolome diversity in 12 divergent algae, an untargeted holistic approach was chosen to cover a broad range of metabolites, such as lipids or pigments known as algal biomarkers. Algal ethyl acetate extracts were analyzed by UHPLC-ESI$^+$-HRMS^2, and acquired ion chromatograms were processed through an untargeted metabolomic workflow in Compound Discoverer 2.1 software (Thermo Scientific) to generate extracted ion chromatograms across samples, detect and quantify corresponding metabolites, and generate the observations/variables matrix used for further statistical analyses. As a first observation, 2565 ions were detected across all samples, and 1143 ions were unique features detected in only one species. Moreover, detected features ranged from 867 ions in *O. tauri* sp. 1 to 241 ions in *Pyramimonas* sp. 1422 ions are shared between at least two species (Figure S1). These observations suggest high variation in the diversity of produced metabolites. To display this variation, an initial principal component analysis (PCA) was conducted on the observations/variables matrix to compare metabolomic profiles between strains at the global metabolome scale (Figure 1). Principal Components (PCs) 1 and 2 describe 34.8% of the variation, and the first PC separates Mamiellales from the two microalgae *P. lutheri* (Haptophyta lineage) and *P. tricornutum* (Stramenopile lineage), hereafter considered outlier microalgae as they are not part of the green lineage. The second PC separates *Ostreococcus* species from other Chlorophytes (*Nephroselmis*, *Pyramimonas*). Interestingly, intraspecific metabolome diversity seems minimal as compared to within-replicate variation, as the two *O. tauri* strains' PCA confidence ellipses overlap.

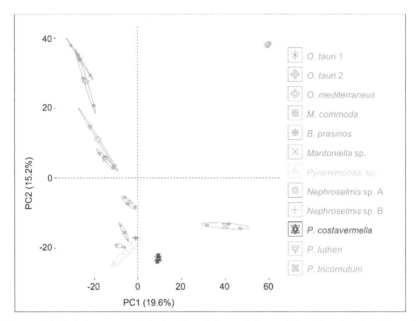

Figure 1. Principal Component Analysis of the whole metabolome of 12 marine microalgae. For each species and strain, confidence ellipses cover 95% of group position estimation.

2.2. Identification of the Major Metabolites and Detection of Chemotaxonomic Markers

In order to go further into the comparative metabolic profiling of the 12 strains, the 10 major metabolites of each strain, defined as the 10 highest peak areas of the extracted ion chromatograms (XIC), were identified (Figure 2). Identification was carried out by comparing compound raw formulas (calculated on the basis of high-resolution mass spectrometry) to databases (Dictionary of Marine Natural Products and SciFinder) to retrieve candidate compounds, then MS^2 spectra were submitted to databases for comparison (Global Natural Products Social Molecular Networking—GNPS [25]) or elucidated to infer putative structures. Compounds were classified into 10 different groups of polar lipids and pigments. Among the polar galactolipids, eight were monogalactosyl diacylglycerols (MGDGs) (Figure S2), two were monogalactosyl monoacylglycerols (MGMGs or Lyso-MGDGs), and one was a sulfoquinovosyl diacylglycerol (SQDG) (Figures S4 and S5).

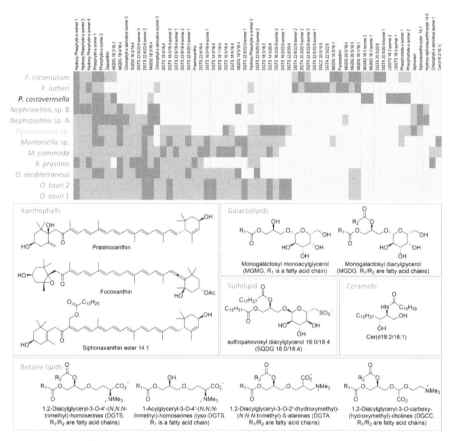

Figure 2. Matrix of the top 10 (blue), detected (green), and not detected (yellow) metabolites, and structure examples of the 59 most abundant compounds over the 12 microalgal strains. For all compounds, identifications are based on (1) molecular formulas, (2) automatic assignment via Global Natural Products Social Molecular Networking (GNPS), (3) interpretation of MS^2 spectra and comparison with published data, and (4) phylogeny.

For MGDGs and SQDGs, the regiochemical assignment (*sn*-1 and *sn*-2 positions) of both fatty acid (FA) chains was done by comparing the MS^2 fragmentation patterns. The fragment resulting from *sn*-1 FA loss ($[M+X-R_1CO_2H]^+$) exhibits a higher peak intensity than the one resulting from *sn*-2 FA loss ($[M+X-R_2CO_2H]^+$) for the protonated adduct (X = H) of SQDGs [26] and sodiated adduct (X = Na) of MGDGs [27]. Besides this, it has been established in the literature that the sugar moiety is a galactose for glycolipids and a sulfoquinovose for SQDG [28]. The betaine lipids are the most represented and diversified group with twenty 1,2-diacylglyceryl-3-*O*-4'-(*N,N,N*-trimethyl)-homoserines (DGTSs) (Figures S6 and S7), seven 1,2-diacylglyceryl-3-*O*-2'-(hydroxymethyl)-(*N,N,N*-trimethyl)-ß-alanines (DGTAs), two 1-acylglyceryl-3-*O*-4'-(*N,N,N*-trimethyl)-homoserines (lyso-DGTSs), and one 1,2-diacylglyceryl-3-*O*-carboxy-(hydroxymethyl)-choline (DGCC) (Figure S8). Regiochemical assignment of FAs was done as for galactolipids on the basis of MS^2 spectra. Here, the collision-induced $[M + Na-R_2CO_2H]^+$ fragment of sodiated adducts produced a higher peak than the $[M+Na-R_1CO_2H]^+$ fragment [29]. Distinction of the isomeric DGTA and DGTS betaine lipids was performed on the basis of strain phylogeny in case of coelution. The fragmentation pattern commonly described for these lipids includes the characteristic 59 Da neutral loss corresponding to the loss of trimethyl amine (NMe$_3$) and the 87 Da neutral loss (CH$_3$-CH$^-$-N$^+$Me$_3$) for DGTSs as a consequence of fragmentation after transposition of the carboxyl group [30]. Unfortunately, this fragmentation reaction was not observed in our analyses. According to the literature, DGTAs are specific to brown algae (*P. lutheri* and *P. tricornutum*) while DGTSs are produced by microalgae from the green lineage [28]. On the pigment side, 11 chlorophylls (Figures S9–S23) and 6 xanthophylls (Figures S24–S30) appeared to be largely shared among strains. In the xanthophyll series, one compound was identified as either prasinoxanthin or its isomer violaxanthin. The uncertainty was eventually disentangled thanks to the typical dehydration fragmentation pattern of prasinoxanthin (Figure S28), while an 80 Da neutral loss was observed for the epoxycarotenoid violaxanthin (Figure S29) [31]. Fucoxanthin (Figure S30) was unambiguously identified from specific fragments at *m/z* 109.1014, 581.3975, and 641.4207 in MS^2 [32]. Apart from the polar lipids and pigments, a ceramide non-hydroxy fatty acid sphingosine (Cer) (Figure S31) was also identified.

The most abundant and diversified metabolites identified over the 12 strains were polar lipids and pigments. These observations are consistent with an increasing number of studies concerning the analysis of algal lipidomes [28,33] and provide new lipidome information for the strains *Mantoniella* sp., *Nephroselmis* sp., and *Pyramimonas* sp. recently isolated from environmental samples. Further phylogenetic signals of metabolites are given and discussed in the following section.

The 10 major metabolites of each microalga shown above were chosen to construct a new matrix of compound abundance to perform a second PCA (Figure 3A). Remarkably, the variability explained by the two PCs remained similar (38.8%), and so did the pattern of clustering of the different strains as compared to the first PCA conducted on the whole metabolome analysis. The first PC again discriminated the outlier microalgae *P. lutheri* and *P. tricornutum* from the Mamiellales, while PC2 separated *P. costavermella* from the brown algae, and the green microalgae fanned out along this axis.

The contribution of each major metabolite to a strain or group of algae can be inferred from the biplot projection of the PCA (Figure 3B–D). The first PC separates the green microalgae from the outlier brown ones, and as expected, this distinction is primarily due to the betaine lipids DGTAs and DGTSs. The high chemical diversity of DGTSs is due to a greater variability in the acyl chain length and number of unsaturations, while DGTAs hold only long (C20–C22) and highly unsaturated acyl chains (Figure 2). MGDGs are represented in every species. MGDGs 18:3/16:4, 18:4/16:4, and 18:5/16:4 are only found in green microalgae. They predominate in the Mamiellales as previously described by Degraeve-Guilbault et al. [34]. Major galactolipids of the brown algae *P. tricornutum* contain 16:0, 16:1, 16:3, and 20:5 fatty acid chains, which is also consistent with previous analyses, reinforcing the reproducibility of these observations [29]. Interestingly, MGMGs are only detected in *P. tricornutum* and *P. costavermella*. Usually, these lipids are not extensively studied in the literature and may be associated with lipid remodeling or environmental plasticity [35]. Some strains can exhibit metabolites exclusive to their group. DGCCs are present in *P. lutheri* but absent in *P. tricornutum* and may be a biomarker of haptophytes [28,36]. This is largely described in the literature, but we also show here that fucoxanthin occurs in the lineage containing *P. lutheri* and *P. tricornutum*. On the other hand, C14 esterified siphonaxanthin (Figures S25–S27) and siphonein (Figure S29) are specific to the Chlorophytes *Nephroselmis* and *Pyramimonas*. In fact, these pigments are widely found in green algae, especially in deep-water or shade species, as they improve the efficiency of light-harvesting complexes [37] or protect the cells from high light damage. Moreover, prasinoxanthin (Figure S31) is the major xanthophyll that identifies Mamiellales amongst other green algae [12].

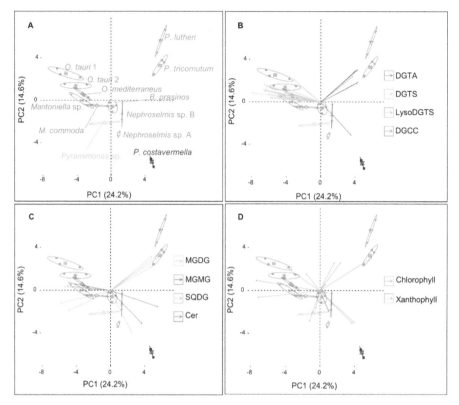

Figure 3. (**A**) Principal component analysis (PCA) constructed from the 59 most abundant metabolites matrix over the 12 microalgal strains, and the corresponding biplots of the (**B**) betaine lipids, (**C**) galacto-, sulfolipids, and ceramide, and (**D**) pigments (each arrow corresponds to a metabolite). Confidence ellipses cover 95% of group position estimation. Arrow coordinates correspond to the contributions of metabolites to the PC and color to the metabolite class. Arrows point toward the strains where they are the most represented.

Beyond a simple identification of metabolite classes within microalgal species or lineages, our results demonstrate that some specific metabolites within classes may serve as phylogenetic markers, pending analysis of additional species identified in each group. Our analysis confirms that prasinoxanthin is specific to the Mamiellales, but also that DGTS 22:6/16:4 is specific to microalgae from this group. The number and diversity of DGTSs we found between *Pyramimonas* sp. and Mamiellales is an outstanding observation, since these betaine lipids were not detected in *Nephroselmis* sp. or even in *B. prasinos* or *O. mediterraneus*. Within the Mamiellales, the *Ostreococcus* genus differs by the presence of two MGDGs (20:5/16:3 and 16:1/16:1). The species *O. tauri* has six DGTSs which are not found in *O. mediterraneus*. Interestingly, the presence/absence matrix does not differentiate both *O. tauri* strains. More data will be necessary to confirm these observations, but the "major metabolites approach" does indeed appear to be interesting for differentiating between these microalgae at the species level.

The choice of taxonomic chemical biomarkers is always challenging as the diversity and abundance of metabolites should reflect species divergence rather than intrinsic variability due to environmental factors. To get around this, chemical classification of plants has been preferentially achieved by comparison of secondary metabolites since these are remarkably diverse [38], include numerous classes of compounds (glycosides, phenolics, or alkaloids) [39], and are relevant for species classification, as they are restricted to taxonomically related groups of species [40,41]. Comparison of algae has so far relied on pigment analysis using 44 pigment types spanning 27 classes of photosynthetic algae [42], and these pigments are consistent with the endosymbiotic evolutionary history of eukaryotes [43]. More recently, many efforts in algal compound screening have enabled the description of hundreds of new metabolites each year [44], which provide the opportunity to identify species from a broader spectrum of compounds. Algal lipids have been extensively described in model species such as *Chlamydomonas reinhardtii*, *Chlorella* sp., *Nannochloropsis* sp., or *P. tricornutum* [29,45–47], while complete lipidome profiles have yet to be acquired for most algae. Nonetheless, lipids, especially FAs [48–50], sterols [50], alkenones [51], or polar lipids [52,53], are widely used as species tracers. It is important to keep in mind that lipid profiles may be impacted by environmental, biotic, or abiotic factors [54], as demonstrated in many studies on nutrient availability, irradiance, and growth stage [29,34,55,56]. However, even though taxonomic signals may be diminished by external factors, it has been shown that taxonomy accounts for 3 to 4 times more variance in the lipid profiles of phytoplankton than abiotic factors [49]. Besides this, polar lipids, especially betaine lipids (DGTA/S, DGCC), constitute the least impacted metabolite class by growth stage as demonstrated by Cañavate and colleagues, and they are therefore considered reliable lipidic taxonomic markers [57]. The abundancy profiles of "major metabolites" are consistent across experiments from available studies performed at the molecular level [29,34,58], suggesting that they are relevant chemotaxonomic marker candidates.

2.3. Phylogenetic Analysis and Metabolome-Based Taxonomy

The phylogeny based on the partial analysis of the 18S rDNA subunit was consistent with previous findings based on a larger set of sequence data [1,59] retracing the molecular divergence between species and the different microalgal families (Figure 4A) and was therefore used as a reference for comparison with the information from the metabolome of each microalga.

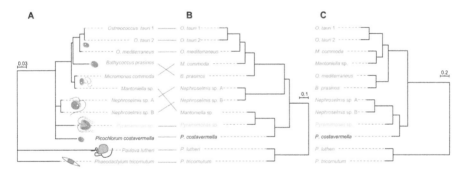

Figure 4. (**A**) Partial 18S ribosomal gene phylogeny based on 429 base pairs; scale indicates substitution per site. (**B**) A hierarchical clustering dendrogram based on 3138 metabolites; the scale represents the Spearman correlation coefficient between strains. (**C**) A hierarchical clustering dendrogram based on the 59 most abundant metabolites; the scale represents the Spearman correlation coefficient between strains.

First, hierarchical clustering analysis (HCA) was performed to determine the metabolomic proximity between strains. The HCA dendrogram (Figure 4B) was generated by calculating the distance matrix between strains based on the Spearman correlation coefficient and aggregated by "complete linkage". A clear clustering of Mamiellales species emerged from HCA, except for *Mantoniella* sp., which exhibited a metabolomic profile more closely related to *Pyramimonas* sp. This peculiarity disappears when the metabolome is reduced to the 10 major metabolites to build the chemotaxonomy (Figure 4C). This suggests that major compounds tend to be reliable species biomarkers when comparing divergent as well as closely related organisms. Microalgal cultures are very rarely axenic, since a bacterial community often co-exists with microalgae [60]. Estimation of the proportion of bacteria by cytometry revealed that most cultures contained less than 5% bacteria, while it turned out that both

Mantoniella sp. and *Pyramimonas* sp. could not be cleansed of bacterial partners (microalgae/bacteria ~1:1). This may be a consequence of their mixotrophic regime [61,62]. The clustering of these strains may thus be due to the metabolomic contribution of bacteria, either because of identical metabolites of bacterial origin or through a similar microalgae–bacteria interaction process that may dramatically influence the metabolome of the microalgal partner. However, we demonstrate here that algal metabolites predominate in the total extract, so bacterial participation disappears when considering the major metabolites only. Besides this, statistical comparison of genetic distances with metabolomic distances based on whole-metabolome analysis amongst algae confirms a strong correlation between both approaches (Mantel test: $r = 0.77$, p-value = 0.001). Thus, there is a good overlap of phylogenetic and chemotaxonomic signals, as metabolomic distances reflect genetic distances between species. This correlation is even stronger when estimated from the most abundant metabolites (Mantel test: $r = 0.90$, p-value = 0.001), confirming the robustness and potential of the metabolomic approach to discriminate and retrace the evolutionary history of divergent species.

Establishing the full lipidome profile of algae at the molecular level within and between species is relevant for several reasons. First, it provides a fundamental description of the species at the metabolite level and helps to estimate the chemical divergence between strains or species. Secondly, it can lead to the identification of algal species producing interesting bioactive compounds, such as high-value carotenoids and lipids, with agronomical and pharmaceutical applications. The present study highlights the abundance of long-chain polyunsaturated FA, such as octadecapentaenoic acid (18:5n-3), eicosapentaenoic acid (20:5n-3), or docosahexaenoic acid (22:6n-3), the health benefits of which have been previously recognized [63], for example, in preventing cardiovascular or mental disorders [64–66]. Moreover, particularly abundant in Mamiellophyceae, polar lipids and especially MGDG 18:3/16:4 and MGDG 18:4/16:4 of *Tetraselmis* sp. have effective anti-inflammatory properties [67,68]. *P. lutheri's* major galactolipid MGDG 20:5/18:4 has also been shown to have anti-inflammatory activity [69], inhibiting both human melanoma cell growth [70] and bacterial development [71]. Last but not least, SQDGs are probably among the most promising compounds in the medical field as lipids from this class are known to act on HIV infection [72], but also exhibit anti-HSV-1 and anti-HSV-2 activities [73] and anti-inflammatory [74] and antitumor properties [75]. The aforementioned bioactive molecules are far from exhaustive but highlight the biotechnological potential of algae as producers of bioactive molecules.

In conclusion, we investigated the metabolomes of 12 microalgal strains from 11 species and characterized the major carotenoids and lipids. The approach of using major metabolites allows microalgal species and lineages to be distinguished. Evolutionary divergence between species can be inferred, in good congruence with the phylogenies obtained from sequence data obtained through a classical molecular approach. Therefore, these results support the hypothesis of a metabolomics equivalent to the "molecular clock" based on the analysis of sequence data. The resulting "metabolomics clock" metaphor is also constrained by the technical challenges raised by the "molecular clock" inferred from DNA sequence analysis: What are the mode and tempo of metabolome evolution? Are some metabolites changing faster than others? What is the distribution of fitness effects of metabolite changes? To answer these questions, statistical developments are needed to develop metabolomic distances and larger datasets including additional species should be obtained to include a broader variation in evolutionary distances.

3. Materials and Methods

3.1. Culture Conditions and Growth Measurement

Cultures were grown in modified Keller Artificial Seawater medium [76] (K-ASWO) containing 420 mM NaCl, 10 mM KCl, 20 mM MgCl$_2$, 10 mM CaCl$_2$, 25 mM MgSO$_4$, 2.5 mM NaHCO$_3$, 0.88 mM NaNO$_3$, 5.0×10^{-5} M NH$_4$Cl, 1.0×10^{-5} M β-glycerophosphate, 1.0×10^{-8} M H$_2$SeO$_3$, 1 mL of 1 M Tris-HCl (pH 7.2) per liter of medium, 3.7×10^{-10} M cyanocobalamin, 2.0×10^{-9} M biotin, and 3.0×10^{-7}

thiamine in addition to Keller trace metal solution [77]. Algal strains were cultured in T75 cell culture flasks with ventilated caps (Sarstedt, Germany) containing 100 mL of K-ASWO medium. Each flask was inoculated to a cell density of 1×10^6 cells.mL^{-1} for *O. tauri* 1 (RCC 6850) and 2 (RCC 4221), *O. mediterraneus* (RCC 2590), *B. prasinos* (RCC 4222), *M. commoda* (RCC 827), *Mantoniella* (RCC 6849), *P. costavermella* (RCC 4223), *P. tricornutum* (RCC 6851), and *P. lutheri* (RCC 6852); 2.4×10^5 cells.mL^{-1} for *Nephroselmis* sp. A (RCC 6846); 3.3×10^5 cells.mL^{-1} for *Nephroselmis* sp. B (RCC 6847); and 7.8×10^4 cells.mL^{-1} for *Pyramimonas* sp. (RCC 6848). Cultures were maintained at a temperature of 20 °C under continuous light of 100 μE.m^{-2}.s^{-1} and were agitated manually once a day; cell density was measured every day by flow cytometry.

3.2. Microalgal Culture Axenization

All cultures were treated with antibiotics to lower the bacterial concentration. Quantities of 50 μg.mL^{-1} ampicillin (A9518, Sigma-Aldrich), 50 μg.mL^{-1} gentamycin (G1914, Sigma-Aldrich), 20 μg.mL^{-1} kanamycin (60615, Sigma-Aldrich), and 100 μg.mL^{-1} neomycin (N6386, Sigma-Aldrich) were added to K-ASWO, and after two subculturing stages, the bacterial content was low enough to perform metabolomic analysis in most strains. A single antibiotic treatment reduced the bacterial contamination in *Ostreococcus* and *P. lutheri* cultures but was unable to remove bacteria completely in others.

3.3. Flow Cytometry

Cells were fixed using glutaraldehyde (0.25% final concentration, G6257, Sigma-Aldrich) with the addition of Pluronic F-68 (0.1% final concentration, P-7061, Sigma-Aldrich) for 15 min in the dark and stained with SYBR Green I (LON50512, Ozyme) for another 15 min in the dark. Cell counting was performed using a Beckman Coulter Cytoflex flow cytometer (laser excitation wavelength 488 nm) by chlorophyll autofluorescence for microalgae (detection filter >620 nm) and by SYBR Green I fluorescence for bacteria (detection bandwidth 525–540 nm, corresponding to the FITC (fluorescein isothiocyanate) channel). Data were analyzed with CytExpert 2.2 software (Beckman Coulter).

3.4. Metabolite Extraction

Microalgal cells were collected three to four days post inoculation (Supplemental Figure S1) by filtration of 100 mL of culture through a Whatman GF/F filter (Z242519, Sigma-Aldrich) under reduced pressure (600 mbar). Then, filters were placed in disposable glass culture tubes with 7 mL of ethyl acetate (16371, Sigma-Aldrich) to solubilize algae cells overnight in a C25 incubator shaker (New Brunswick Scientific, 100 rpm, 19 °C).

3.5. UHPLC-HRMS Analyses

Microalgal extracts were analyzed on an Ultimate 3000 UHPLC Dionex system coupled to an Orbitrap MS/MS FT Q-Exactive Focus Thermo Scientific mass spectrometer. Samples were solubilized in MeOH (1 mg.mL^{-1}) and 1 μL was injected onto the column. The column was a Phenomenex Luna Omega Polar C18 (150 × 2.1 mm, 1.6 μm, 100 Å) conditioned at 42 °C. The mobile phase was a mixture of water (solvent A) with increasing proportion of acetonitrile (solvent B, 012041, Biosolve), both solvents modified with 0.1% of formic acid. The gradient was as follows: 50% B from 3 min before injection to 1 min after; between 1 and 3 min, a linear increase of B up to 85%, followed by 85% B for 2 min; 89% B from 5.1 to 7 min; 93% B from 7.1 to 10 min; 97% B from 10.1 to 13 min; and finally, 100% B from 13.1 to 18 min. The flow was set to 0.5 mL.min^{-1} and injected into the mass spectrometer 1 min after injection (diverted before). Mass spectrometry analyses were performed in the positive electrospray ionization mode in the 133.4–2000 Da range, and mass spectra were recorded in the centroid mode. The mass spectrometer method was set to FullMS data-dependent MS^2. In fullMS, the resolution was set to 70,000 and the AGC target to 3×10^6 for a chromatogram peak width (FWHM) of 6 s. In MS^2, the resolution was 17,500, the AGC target 1×10^5, the isolation window 0.4 Da, and the stepped normalized collision energy 15/30/45 with 10 s of dynamic exclusion. The lock mass was calibrated on the $Cu(CH_3CN)^{2+}$ ion at *m/z* 144.9821 Da.

3.6. LC-MS Data Preprocessing

Total ion chromatograms were processed through the untargeted metabolomic workflow of Compound Discoverer (CD) 2.1 (Thermo Scientific). A Quality Control mix (QC) composed of the 12 algal extracts was analyzed together with algal extracts and K-ASWO medium used as a blank to remove nonalgal compounds. The CD workflow performs retention time correction, detection of unknown compounds, and grouping across samples; fills gaps when features are absent; hides chemical background (using blank samples); and finally predicts compound elemental composition. The retention time window was set to 2–18 min. The maximum time shift for compound alignment was 0.1 min, the maximum mass tolerance for compound grouping and elemental composition calculation was 3 ppm, and the minimum peak intensity was 2×10^6. This workflow provided an observation/variable matrix used for further statistical analysis.

3.7. Confirmation of Algae Identities and Reconstruction of Phylogenies

Algae identification was performed on the basis of partial 18S rDNA sequence analysis. Total DNA was extracted with hexadecyltrimethylammonium bromide (CTAB) as described by Winnepenninckx et al. [78]. The 18S rDNA gene region was amplified by PCR using the unique nondegenerate universal eukaryotic forward primer F-566 (5′ CAG CAG CCG CGG TAA TTC C 3′) and the reverse primer R-1200 (3′ CCC GTG TTG AGT CAA ATT AAG C 5′) [79] before sequencing by the GATC company.

Then, partial 18S rDNA sequences were aligned using MUSCLE 3.8 [80], and gaps were manually removed to get 429 base pairs sequences. Best substitution model selection and phylogenetic tree reconstruction was performed using IQ-TREE 1.6.12. The best selected model was TIM2e+G4 with the Bayesian Inference Criterion, and 1000 bootstraps were used to construct the consensus phylogenetic tree.

3.8. Figure Plotting and Statistical Analysis

All figures were plotted using R 3.6.1. Principal Component Analysis and corresponding biplots were calculated and constructed using the FactoMineR v1.42 package and PCA function with the scaled data option and 95% confidence ellipse lines. The phylogenetic tree was plotted using the phytool v0.6-99 package, and the patristic matrix was calculated using the ape v5.3 package, while the metabolite distance correlation matrix was calculated using the R base stats functions. The Mantel test was performed using the ade4 v1.7-4 package.

Supplementary Materials: The following are available online at http://www.mdpi.com/1660-3397/18/2/78/s1, Figure S1: Histogram of detected compounds (A) and occurrence in algal strains (B), Figure S2: Microalgal concentration growth curves over time, Table S1: Algal and bacterial concentration (cells.mL^{-1}) at sampling day for the 12 algae species, Figures S3–S31: Analytical data on microalgal metabolites.

Author Contributions: Conceptualization, G.P. and D.S.; methodology, R.M.-G., G.P., D.S.; data curation, R.M.-G.; writing—original draft preparation, R.M.-G., G.P., D.S.; writing—review and editing, R.M.-G., G.P., D.S.; supervision, G.P., D.S.; project administration, G.P., D.S.; funding acquisition, G.P., D.S. All authors have read and agreed to the published version of the manuscript.

Funding: This project obtained funding from the Interaction in Marine Organisms initiative of the FR3724 (Observatoire Océanologique de Banyuls-sur-Mer).

Acknowledgments: We would like to thank Manon Norest and Adrien Cadoudal for their help with strain isolation and molecular sequencing, the BIO2MAR platform for help with metabolomic analyses, the BIOPIC platform for help with cytometry, and the Genophy team members for support and stimulating discussions on this project. Special thanks to Valérie Domien for help with production of the microalgal graphics and Nigel Grimsley for English language corrections.

Conflicts of Interest: The authors declare no conflict of interest.

References

1. Not, F.; Siano, R.; Kooistra, W.H.C.F.; Simon, N.; Vaulot, D.; Probert, I. Diversity and ecology of eukaryotic marine phytoplankton. In *Advances in Botanical Research*; Elsevier: Amsterdam, The Netherlands, 2012; Volume 64, pp. 1–53. ISBN 978-0-12-391499-6.
2. Burki, F.; Roger, A.J.; Brown, M.W.; Simpson, A.G.B. The new tree of eukaryotes. *Trends Ecol. Evol.* **2020**, *35*, 43–55. [CrossRef]
3. Leliaert, F.; Verbruggen, H.; Zechman, F.W. Into the deep: New discoveries at the base of the green plant phylogeny. *BioEssays* **2011**, *33*, 683–692. [CrossRef]
4. Gould, S.B.; Waller, R.F.; McFadden, G.I. Plastid evolution. *Annu. Rev. Plant Biol.* **2008**, *59*, 491–517. [CrossRef] [PubMed]
5. Lughadha, E.N.; Govaerts, R.; Belyaeva, I.; Black, N.; Lindon, H.; Allkin, R.; Magill, R.E.; Nicolson, N. Counting counts: Revised estimates of numbers of accepted species of flowering plants, seed plants, vascular plants and land plants with a review of other recent estimates. *Phytotaxa* **2016**, *272*, 82. [CrossRef]
6. Tragin, M.; Vaulot, D. Novel diversity within marine *Mamiellophyceae* (Chlorophyta) unveiled by metabarcoding. *Sci. Rep.* **2019**, *9*, 1–14. [CrossRef] [PubMed]
7. Massana, R. Eukaryotic picoplankton in surface oceans. *Annu. Rev. Microbiol.* **2011**, *65*, 91–110. [CrossRef]
8. Courties, C.; Vaquer, A.; Troussellier, M.; Lautier, J.; Chrétiennot-Dinet, M.J.; Neveux, J.; Machado, C.; Claustre, H. Smallest eukaryotic organism. *Nature* **1994**, *370*, 255. [CrossRef]
9. Schaum, E.; Rost, B.; Millar, A.J.; Collins, S. Variation in plastic responses of a globally distributed picoplankton species to ocean acidification. *Nat. Clim. Chang.* **2013**, *3*, 298–302. [CrossRef]
10. Lang, D.; Weiche, B.; Timmerhaus, G.; Richardt, S.; Riaño-Pachón, D.M.; Corrêa, L.G.G.; Reski, R.; Mueller-Roeber, B.; Rensing, S.A. Genome-wide phylogenetic comparative analysis of plant transcriptional regulation: A timeline of loss, gain, expansion, and correlation with complexity. *Genome Biol. Evol.* **2010**, *2*, 488–503. [CrossRef]
11. Guiry, M.D.; Guiry, G.M.; Morrison, L.; Rindi, F.; Miranda, S.V.; Mathieson, A.C.; Parker, B.C.; Langangen, A.; John, D.M.; Bárbara, I.; et al. AlgaeBase: An on-line resource for algae. *Cryptogam. Algol.* **2014**, *35*, 105–115. [CrossRef]
12. Foss, P.; Guillard, R.R.L.; Liaaen-Jensen, S. Prasinoxanthin—A chemosystematic marker for algae. *Phytochemistry* **1984**, *23*, 1629–1633. [CrossRef]
13. Latasa, M.; Scharek, R.; Gall, F.L.; Guillou, L. Pigment suites and taxonomic groups in Prasinophyceae. *J. Phycol.* **2004**, *40*, 1149–1155. [CrossRef]
14. Serive, B.; Nicolau, E.; Bérard, J.-B.; Kaas, R.; Pasquet, V.; Picot, L.; Cadoret, J.-P. Community analysis of pigment patterns from 37 microalgae strains reveals new carotenoids and porphyrins characteristic of distinct strains and taxonomic groups. *PLoS ONE* **2017**, *12*, e0171872. [CrossRef]
15. Conway, M.; Mulhern, M.; McSorley, E.; van Wijngaarden, E.; Strain, J.; Myers, G.; Davidson, P.; Shamlaye, C.; Yeates, A. Dietary determinants of polyunsaturated fatty acid (PUFA) status in a high fish-eating cohort during pregnancy. *Nutrients* **2018**, *10*, 927. [CrossRef] [PubMed]
16. Abel, S.; Riedel, S.; Gelderblom, W.C.A. Dietary PUFA and cancer. *Proc. Nutr. Soc.* **2014**, *73*, 361–367. [CrossRef] [PubMed]
17. Abdelhamid, A.S.; Martin, N.; Bridges, C.; Brainard, J.S.; Wang, X.; Brown, T.J.; Hanson, S.; Jimoh, O.F.; Ajabnoor, S.M.; Deane, K.H.; et al. Polyunsaturated fatty acids for the primary and secondary prevention of cardiovascular disease. *Cochrane Database Syst. Rev.* **2018**. [CrossRef]
18. Bruno, A.; Rossi, C.; Marcolongo, G.; Di Lena, A.; Venzo, A.; Berrie, C.P.; Corda, D. Selective in vivo anti-inflammatory action of the galactolipid monogalactosyldiacylglycerol. *Eur. J. Pharmacol.* **2005**, *524*, 159–168. [CrossRef]
19. Wang, H.; Li, Y.-L.; Shen, W.-Z.; Rui, W.; Ma, X.-J.; Cen, Y.-Z. Antiviral activity of a sulfoquinovosyldiacylglycerol (SQDG) compound isolated from the green alga *Caulerpa racemosa*. *Bot. Mar.* **2007**, *50*, 185–190. [CrossRef]
20. Blanc-Mathieu, R.; Verhelst, B.; Derelle, E.; Rombauts, S.; Bouget, F.-Y.; Carré, I.; Château, A.; Eyre-Walker, A.; Grimsley, N.; Moreau, H.; et al. An improved genome of the model marine alga *Ostreococcus tauri* unfolds by assessing Illumina de novo assemblies. *BMC Genom.* **2014**, *15*, 1103. [CrossRef]

21. Yau, S.; Krasovec, M.; Benites, L.F.; Rombauts, S.; Groussin, M.; Vancaester, E.; Aury, J.-M.; Derelle, E.; Desdevises, Y.; Escande, M.-L.; et al. Virus-host coexistence in phytoplankton through the genomic lens. *Sci. Adv.* **2020**, in press.

22. Moreau, H.; Verhelst, B.; Couloux, A.; Derelle, E.; Rombauts, S.; Grimsley, N.; Van Bel, M.; Poulain, J.; Katinka, M.; Hohmann-Marriott, M.F.; et al. Gene functionalities and genome structure in *Bathycoccus prasinos* reflect cellular specializations at the base of the green lineage. *Genome Biol.* **2012**, *13*, R74. [CrossRef] [PubMed]

23. Simon, N.; Foulon, E.; Grulois, D.; Six, C.; Desdevises, Y.; Latimier, M.; Le Gall, F.; Tragin, M.; Houdan, A.; Derelle, E.; et al. Revision of the genus *Micromonas* Manton et Parke (*Chlorophyta, Mamiellophyceae*), of the type species *M. pusilla* (Butcher) Manton & Parke and of the species *M. commoda* van Baren, Bachy and Worden and description of two new species based on the genetic and phenotypic characterization of cultured isolates. *Protist* **2017**, *168*, 612–635. [PubMed]

24. Krasovec, M.; Vancaester, E.; Rombauts, S.; Bucchini, F.; Yau, S.; Hemon, C.; Lebredonchel, H.; Grimsley, N.; Moreau, H.; Sanchez-Brosseau, S.; et al. Genome analyses of the microalga *Picochlorum* provide insights into the evolution of thermotolerance in the green lineage. *Genome Biol. Evol.* **2018**, *10*, 2347–2365. [CrossRef] [PubMed]

25. Wang, M.; Carver, J.J.; Phelan, V.V.; Sanchez, L.M.; Garg, N.; Peng, Y.; Nguyen, D.D.; Watrous, J.; Kapono, C.A.; Luzzatto-Knaan, T.; et al. Sharing and community curation of mass spectrometry data with Global Natural Products Social Molecular Networking. *Nat. Biotechnol.* **2016**, *34*, 828–837. [CrossRef] [PubMed]

26. Zianni, R.; Bianco, G.; Lelario, F.; Losito, I.; Palmisano, F.; Cataldi, T.R.I. Fatty acid neutral losses observed in tandem mass spectrometry with collision-induced dissociation allows regiochemical assignment of sulfoquinovosyl-diacylglycerols: The neutral loss of FAs from SQDGs by tandem MS. *J. Mass Spectrom.* **2013**, *48*, 205–215. [CrossRef] [PubMed]

27. Guella, G.; Frassanito, R.; Mancini, I. A new solution for an old problem: The regiochemical distribution of the acyl chains in galactolipids can be established by electrospray ionization tandem mass spectrometry. *Rapid Commun. Mass Spectrom.* **2003**, *17*, 1982–1994. [CrossRef]

28. Mimouni, V.; Couzinet-Mossion, A.; Ulmann, L.; Wielgosz-Collin, G. Lipids from microalgae. In *Microalgae in Health and Disease Prevention*; Elsevier: Amsterdam, The Netherlands, 2018; pp. 109–131. ISBN 978-0-12-811405-6.

29. Abida, H.; Dolch, L.-J.; Meï, C.; Villanova, V.; Conte, M.; Block, M.A.; Finazzi, G.; Bastien, O.; Tirichine, L.; Bowler, C.; et al. Membrane glycerolipid remodeling triggered by nitrogen and phosphorus starvation in *Phaeodactylum tricornutum*. *Plant Physiol.* **2015**, *167*, 118–136. [CrossRef]

30. Roche, S.A.; Leblond, J.D. Betaine lipids in chlorarachniophytes. *Phycol. Res.* **2010**, *58*, 298–305. [CrossRef]

31. Rivera, S.M.; Christou, P.; Canela-Garayoa, R. Identification of carotenoids using mass spectrometry. *Mass Spectrom. Rev.* **2014**, *33*, 353–372. [CrossRef]

32. Zhang, Y.; Wu, H.; Wen, H.; Fang, H.; Hong, Z.; Yi, R.; Liu, R. Simultaneous determination of *fucoxanthin* and its deacetylated metabolite fucoxanthinol in rat plasma by Liquid Chromatography-Tandem Mass Spectrometry. *Mar. Drugs* **2015**, *13*, 6521–6536. [CrossRef]

33. Harwood, J.L.; Guschina, I.A. The versatility of algae and their lipid metabolism. *Biochimie* **2009**, *91*, 679–684. [CrossRef] [PubMed]

34. Degraeve-Guilbault, C.; Bréhélin, C.; Haslam, R.; Sayanova, O.; Marie-Luce, G.; Jouhet, J.; Corellou, F. Glycerolipid Characterization and nutrient deprivation-associated changes in the green picoalga *Ostreococcus tauri*. *Plant Physiol.* **2017**, *173*, 2060–2080. [CrossRef] [PubMed]

35. Da Costa, E.; Domingues, P.; Melo, T.; Coelho, E.; Pereira, R.; Calado, R.; Abreu, M.H.; Domingues, M.R. Lipidomic signatures reveal seasonal shifts on the relative abundance of high-valued lipids from the brown algae *Fucus vesiculosus*. *Mar. Drugs* **2019**, *17*, 335. [CrossRef] [PubMed]

36. Kato, M.; Sakai, M.; Adachi, K.; Ikemoto, H.; Sano, H. Distribution of betaine lipids in marine algae. *Phytochemistry* **1996**, *42*, 1341–1345. [CrossRef]

37. Akimoto, S.; Tomo, T.; Naitoh, Y.; Otomo, A.; Murakami, A.; Mimuro, M. Identification of a new excited state responsible for the in vivo unique absorption band of siphonaxanthin in the green alga *Codium fragile*. *J. Phys. Chem. B* **2007**, *111*, 9179–9181. [CrossRef]

38. Kessler, A.; Kalske, A. Plant secondary metabolite diversity and species interactions. *Annu. Rev. Ecol. Evol. Syst.* **2018**, *49*, 115–138. [CrossRef]

39. Wink, M. Introduction: Biochemistry, physiology and ecological functions of secondary metabolites. In *Biochemistry of Plant Secondary Metabolism*; Wink, M., Ed.; Wiley-Blackwell: Oxford, UK, 2010; pp. 1–19. ISBN 978-1-4443-2050-3.

40. Singh, R. Chemotaxonomy: A tool for plant classification. *J. Med. Plants Stud.* **2016**, *4*, 90–93.

41. Wink, M.; Botschen, F.; Gosmann, C.; Schfer, H.; Waterman, P.G. Chemotaxonomy seen from a phylogenetic perspective and evolution of secondary metabolism. In *Biochemistry of Plant Secondary Metabolism*; Wink, M., Ed.; Wiley-Blackwell: Oxford, UK, 2010; pp. 364–433. ISBN 978-1-4443-2050-3.

42. Mc Gee, D.; Gillespie, E. The bioactivity and chemotaxonomy of microalgal carotenoids. In *Biodiversity and Chemotaxonomy*; Ramawat, K.G., Ed.; Springer International Publishing: Cham, Switerland, 2019; Volume 24, pp. 215–237. ISBN 978-3-030-30745-5.

43. Zimorski, V.; Ku, C.; Martin, W.F.; Gould, S.B. Endosymbiotic theory for organelle origins. *Curr. Opin. Microbiol.* **2014**, *22*, 38–48. [CrossRef]

44. Carroll, A.R.; Copp, B.R.; Davis, R.A.; Keyzers, R.A.; Prinsep, M.R. Marine natural products. *Nat. Prod. Rep.* **2019**, *36*, 122–173. [CrossRef]

45. Martin, G.J.O.; Hill, D.R.A.; Olmstead, I.L.D.; Bergamin, A.; Shears, M.J.; Dias, D.A.; Kentish, S.E.; Scales, P.J.; Botté, C.Y.; Callahan, D.L. Lipid profile remodeling in response to nitrogen deprivation in the microalgae *Chlorella* sp. (Trebouxiophyceae) and *Nannochloropsis* sp. (Eustigmatophyceae). *PLoS ONE* **2014**, *9*, e103389. [CrossRef]

46. Siaut, M.; Cuiné, S.; Cagnon, C.; Fessler, B.; Nguyen, M.; Carrier, P.; Beyly, A.; Beisson, F.; Triantaphylidès, C.; Li-Beisson, Y.; et al. Oil accumulation in the model green alga *Chlamydomonas reinhardtii*: Characterization, variability between common laboratory strains and relationship with starch reserves. *BMC Biotechnol.* **2011**, *11*, 7. [CrossRef]

47. Vieler, A.; Brubaker, S.B.; Vick, B.; Benning, C.A. Lipid droplet protein of *Nannochloropsis* with functions partially analogous to plant oleosins. *Plant Physiol.* **2012**, *158*, 1562–1569. [CrossRef]

48. Lang, I.; Hodac, L.; Friedl, T.; Feussner, I. Fatty acid profiles and their distribution patterns in microalgae: A comprehensive analysis of more than 2000 strains from the SAG culture collection. *BMC Plant Biol.* **2011**, *11*, 124. [CrossRef]

49. Galloway, A.W.E.; Winder, M. Partitioning the relative importance of phylogeny and environmental conditions on phytoplankton fatty acids. *PLoS ONE* **2015**, *10*, e0130053. [CrossRef]

50. Taipale, S.J.; Hiltunen, M.; Vuorio, K.; Peltomaa, E. Suitability of phytosterols alongside fatty acids as chemotaxonomic biomarkers for phytoplankton. *Front. Plant Sci.* **2016**, *7*, 212. [CrossRef]

51. Wolhowe, M.D.; Prahl, F.G.; White, A.E.; Popp, B.N.; Rosas-Navarro, A. A biomarker perspective on coccolithophorid growth and export in a stratified sea. *Prog. Oceanogr.* **2014**, *122*, 65–76. [CrossRef]

52. Van Mooy, B.A.S.; Fredricks, H.F. Bacterial and eukaryotic intact polar lipids in the eastern subtropical South Pacific: Water-column distribution, planktonic sources, and fatty acid composition. *Geochim. Cosmochim. Acta* **2010**, *74*, 6499–6516. [CrossRef]

53. Cañavate, J.P.; Armada, I.; Ríos, J.L.; Hachero-Cruzado, I. Exploring occurrence and molecular diversity of betaine lipids across taxonomy of marine microalgae. *Phytochemistry* **2016**, *124*, 68–78. [CrossRef]

54. Guschina, I.A.; Harwood, J.L. Algal lipids and effect of the environment on their biochemistry. In *Lipids in Aquatic Ecosystems*; Kainz, M., Brett, M.T., Arts, M.T., Eds.; Springer: New York, NY, USA, 2009; pp. 1–24. ISBN 978-0-387-88607-7.

55. Alonso, D.L.; Belarbi, E.-H.; Fernández-Sevilla, J.M.; Rodríguez-Ruiz, J.; Grima, E.M. Acyl lipid composition variation related to culture age and nitrogen concentration in continuous culture of the microalga *Phaeodactylum tricornutum*. *Phytochemistry* **2000**, *54*, 461–471. [CrossRef]

56. Abo-State, M.A.M.; Shanab, S.M.M.; Ali, H.E.A. Effect of nutrients and gamma radiation on growth and lipid accumulation of *Chlorella vulgaris* for biodiesel production. *J. Radiat. Res. Appl. Sci.* **2019**, *12*, 332–342. [CrossRef]

57. Cañavate, J.P.; Armada, I.; Hachero-Cruzado, I. Polar lipids analysis of cultured phytoplankton reveals significant inter-taxa changes, low influence of growth stage, and usefulness in chemotaxonomy. *Microb. Ecol.* **2017**, *73*, 755–774. [CrossRef] [PubMed]

58. Tsugawa, H.; Satoh, A.; Uchino, H.; Cajka, T.; Arita, M.; Arita, M. Mass spectrometry data repository enhances novel metabolite discoveries with advances in computational metabolomics. *Metabolites* **2019**, *9*, 119. [CrossRef]

59. Marin, B.; Melkonian, M. Molecular phylogeny and classification of the Mamiellophyceae class. nov. (Chlorophyta) based on sequence comparisons of the nuclear- and plastid-encoded rRNA operons. *Protist* **2010**, *161*, 304–336. [CrossRef]

60. Abby, S.S.; Touchon, M.; De Jode, A.; Grimsley, N.; Piganeau, G. Bacteria in *Ostreococcus tauri* cultures—friends, foes or hitchhikers? *Front. Microbiol.* **2014**, *5*, 505. [CrossRef]

61. Gast, R.J.; McKie-Krisberg, Z.M.; Fay, S.A.; Rose, J.M.; Sanders, R.W. Antarctic mixotrophic protist abundances by microscopy and molecular methods. *FEMS Microbiol. Ecol.* **2014**, *89*, 388–401. [CrossRef] [PubMed]

62. Anderson, R.; Jürgens, K.; Hansen, P.J. Mixotrophic Phytoflagellate Bacterivory Field measurements strongly biased by standard approaches: A case study. *Front. Microbiol.* **2017**, *8*. [CrossRef] [PubMed]

63. Burri, L.; Hoem, N.; Banni, S.; Berge, K. Marine omega-3 phospholipids: Metabolism and biological activities. *Int. J. Mol. Sci.* **2012**, *13*, 15401–15419. [CrossRef] [PubMed]

64. Simopoulos, A.P. The importance of the ratio of omega-6/omega-3 essential fatty acids. *Biomed. Pharmacother.* **2002**, *56*, 365–379. [CrossRef]

65. Cardozo, K.H.M.; Guaratini, T.; Barros, M.P.; Falcão, V.R.; Tonon, A.P.; Lopes, N.P.; Campos, S.; Torres, M.A.; Souza, A.O.; Colepicolo, P.; et al. Metabolites from algae with economical impact. *Comp. Biochem. Physiol. Part C Toxicol. Pharmacol.* **2007**, *146*, 60–78. [CrossRef]

66. Bowen, K.J.; Harris, W.S.; Kris-Etherton, P.M. Omega-3 fatty acids and cardiovascular disease: Are there benefits? *Curr. Treat. Options Cardiovasc. Med.* **2016**, *18*, 69. [CrossRef]

67. Banskota, A.H.; Stefanova, R.; Gallant, P.; McGinn, P.J. Mono- and digalactosyldiacylglycerols: Potent nitric oxide inhibitors from the marine microalga *Nannochloropsis granulata*. *J. Appl. Phycol.* **2013**, *25*, 349–357. [CrossRef]

68. Banskota, A.H.; Stefanova, R.; Sperker, S.; Lall, S.; Craigie, J.S.; Hafting, J.T. Lipids isolated from the cultivated red alga *Chondrus crispus* inhibit nitric oxide production. *J. Appl. Phycol.* **2014**, *26*, 1565–1571. [CrossRef]

69. Lopes, G.; Daletos, G.; Proksch, P.; Andrade, P.; Valentão, P. Anti-inflammatory potential of monogalactosyl diacylglycerols and a monoacylglycerol from the edible brown seaweed *Fucus spiralis* Linnaeus. *Mar. Drugs* **2014**, *12*, 1406–1418. [CrossRef] [PubMed]

70. Plouguerné, E.; da Gama, B.A.P.; Pereira, R.C.; Barreto-Bergter, E. Glycolipids from seaweeds and their potential biotechnological applications. *Front. Cell. Infect. Microbiol.* **2014**, *4*, 174. [CrossRef]

71. Kim, Y.H.; Kim, E.-H.; Lee, C.; Kim, M.-H.; Rho, J.-R. Two new monogalactosyl diacylglycerols from brown alga *Sargassum thunbergii*. *Lipids* **2007**, *42*, 395–399. [CrossRef] [PubMed]

72. Gustafson, K.R.; Cardellina, J.H.; Fuller, R.W.; Weislow, O.S.; Kiser, R.F.; Snader, K.M.; Patterson, G.M.L.; Boyd, M.R. AIDS-antiviral sulfolipids from cyanobacteria (blue-green algae). *JNCI J. Natl. Cancer Inst.* **1989**, *81*, 1254–1258. [CrossRef]

73. De Souza, L.M.; Sassaki, G.L.; Romanos, M.T.V.; Barreto-Bergter, E. Structural characterization and anti-HSV-1 and HSV-2 activity of glycolipids from the marine algae *Osmundaria obtusiloba* isolated from southeastern Brazilian coast. *Mar. Drugs* **2012**, *10*, 918–931. [CrossRef]

74. Morimoto, T.; Murakami, N.; Nagatsu, A.; Sakakibara, J. Studies on glycolipids. VII. Isolation of two new sulfoquinovosyl diacylglycerols from the green alga *Chlorella vulgaris*. *Chem. Pharm. Bull.* **1993**, *41*, 1545–1548. [CrossRef]

75. Murakami, C.; Kumagai, T.; Hada, T.; Kanekazu, U.; Nakazawa, S.; Kamisuki, S.; Maeda, N.; Xu, X.; Yoshida, H.; Sugawara, F.; et al. Effects of glycolipids from spinach on mammalian DNA polymerases. *Biochem. Pharmacol.* **2003**, *65*, 259–267. [CrossRef]

76. Djouani-Tahri, E.B.; Sanchez, F.; Lozano, J.-C.; Bouget, F.-Y. A phosphate-regulated promoter for fine-tuned and reversible overexpression in *Ostreococcus*: Application to circadian clock functional analysis. *PLoS ONE* **2011**, *6*, e28471. [CrossRef]

77. Keller, M.D.; Selvin, R.C.; Claus, W.; Guillard, R.R.L. Media for the culture of oceanic ultraphytoplankton. *J. Phycol.* **2007**, *23*, 633–638. [CrossRef]

78. Winnepenninckx, B. Extraction of high molecular weight DNA from molluscs. *Trends Genet.* **1993**, *9*, 407. [PubMed]

79. Hadziavdic, K.; Lekang, K.; Lanzen, A.; Jonassen, I.; Thompson, E.M.; Troedsson, C. Characterization of the 18S rRNA gene for designing universal eukaryote specific primers. *PLoS ONE* **2014**, *9*, e87624. [CrossRef] [PubMed]

80. Madeira, F.; Park, Y.M.; Lee, J.; Buso, N.; Gur, T.; Madhusoodanan, N.; Basutkar, P.; Tivey, A.R.N.; Potter, S.C.; Finn, R.D.; et al. The EMBL-EBI search and sequence analysis tools APIs in 2019. *Nucleic Acids Res.* **2019**, *47*, W636–W641. [CrossRef]

 *marine drugs*

Article

Four New Isocoumarins and a New Natural Tryptamine with Antifungal Activities from a Mangrove Endophytic Fungus *Botryosphaeria ramosa* L29

Zhihui Wu [1], Jiaqing Chen [1], Xiaolin Zhang [1], Zelin Chen [1], Tong Li [1], Zhigang She [2], Weijia Ding [1,*] and Chunyuan Li [1,*]

[1] College of Materials and Energy, South China Agricultural University, Guangzhou 510642, China; w_zhi_hui@sina.com (Z.W.); ch_jiaqing@sina.com (J.C.); catherinezxll@sina.com (X.Z.); czrin@sina.cn (Z.C.); zjjlitong@sina.com (T.L.)
[2] School of Chemistry, Sun Yat-Sen University, Guangzhou 510275, China; cesshzhg@mail.sysu.edu.cn
* Correspondence: dwjzsu@163.com (W.D.); chunyuanli@scau.edu.cn (C.L.); Tel.: +86-020-8528-0319 (W.D. & C.L.)

Received: 8 January 2019; Accepted: 22 January 2019; Published: 1 February 2019

Abstract: Four new isocoumarin derivatives, botryospyrones A (**1**), B (**2**), C (**3**), and D (**4**), and a new natural tryptamine, (3a*S*, 8a*S*)-1-acetyl-1, 2, 3, 3a, 8, 8a-hexahydropyrrolo [2,3b] indol-3a-ol (**5**), were isolated from a marine mangrove endophytic fungus *Botryosphaeria ramosa* L29, obtained from the leaf of *Myoporum bontioides*. Their structures were elucidated using spectroscopic analysis. The absolute configurations of compounds **3**, **4**, and **5** were determined by comparison of their circular dichroism (CD) spectra with the calculated data. The inhibitory activities of compound **1** on *Fusarium oxysporum*, of compounds **2** and **3** on *F. oxysporum* and *Fusarium graminearum*, and of compound **5** on *F. oxysporum*, *Penicillium italicum*, and *F. graminearum* were higher than those of triadimefon, widely used as an agricultural fungicide. Compound **5** was produced after using the strategy we called "using inhibitory stress from components of the host" (UISCH), wherein (2*R*, 3*R*)-3, 5, 7-trihydroxyflavanone 3-acetate, a component of *M. bontioides* with weak growth inhibitory activity towards *B. ramosa* L29, was introduced into the culture medium.

Keywords: isocoumarin; tryptamine; *Botryosphaeria ramose*; antifungal activity

1. Introduction

Since the 1990s, marine-derived fungi including mangrove fungi have attracted considerable interest as a target source of bioactive natural products with chemodiversity [1–3]. *Myoporum bontioides* (Siebold and Zucc.) A. Gray (Scrophulariaceae) is one of the mangrove plants distributed along the coasts of Asian countries [4]. Our chemical investigations into this species have led to the discovery of various flavonoids, sesquiterpenoids, and phenylethanoids, among others [5]. Through our ongoing search for novel bioactive metabolites from the endophytic fungi of this plant [6–9], we have investigated the secondary metabolites of the fungus *Botryosphaeria ramosa* L29, the ethanolic extract of which was found to possess antifungal activity. Four new isocoumarin derivatives, named botryospyrones A (**1**), B (**2**), C (**3**), and D (**4**) (Figure 1) were isolated. However, under such a common cultural condition, the fungal strain faced no living stress from the environment. A plausible question is whether inhibitory stressors from components of the host under natural conditions might change the metabolites of the fungal strain. To provide an answer to this question, a strategy which we called "using inhibitory stress from components of the host" (UISCH) was used and prompted the endophyte to produce the novel natural product (3a*S*, 8a*S*)-1-acetyl-1, 2, 3, 3a, 8, 8a-hexahydropyrrolo [2,3b] indol-3a-ol (**5**) belonging to the tryptamine class.

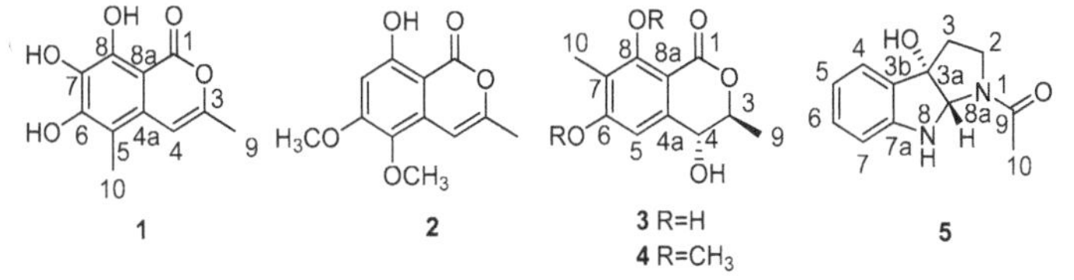

Figure 1. Chemical structures of the isolated compounds **1–5**.

Herein, we report the details of the isolation, structure elucidation, and antifungal activity of these compounds.

2. Results and Discussion

Compounds **1–4** were obtained from the fungus *Botryosphaeria ramosa* L29 grown in common autoclaved rice medium. Compound **5** was produced when using the UISCH strategy. The flavonoid (2*R*, 3*R*)-3, 5, 7-trihydroxyflavanone 3-acetate, a component of *M. bontioides* [10] which had been found by us to have 18.56% growth inhibition rate (0.25 mM) to the mycelium of the fungus *Botryosphaeria ramosa* L29, was added to the autoclaved rice medium. Under these conditions, compound **5** was generated and purified.

Compound **1** was isolated as a yellow powder. The high resolution electron spray ionization mass spectroscopy (HRESIMS) spectrum of **1** showed an ion peak at m/z 223.0608 ([M + H]$^+$, calcd. for $C_{11}H_{11}O_5$ 223.0606), corresponding to the molecular formula $C_{11}H_{10}O_5$. The infrared radiation (IR) bands of **1** at 3244, 1682, 1622, and 1615 cm^{-1} represented the hydroxyl, ester carbonyl, and aromatic ring groups, respectively. The ^{13}C NMR (Table 1) and heteronuclear singular quantum correlation (HSQC) spectrum of **1** showed 11 carbon signals consisting of two methyl groups, one sp^2 methine group, seven olefinic quaternary carbons, and one ester carbonyl carbon. The ^{1}H NMR (Table 1) and HSQC spectra, along with the molecular formula of **1** indicated signals for one hydrogen-bond hydroxyl group (δ_H 11.18), two aromatic hydroxyl groups (δ_H 5.38 and 6.42), one olefinic proton (δ_H 6.38), and two methyl groups (δ_H 2.18 and 2.29). These characteristics suggested that compound **1** might be an isocoumarin [11] substituted with three hydroxyl and two methyl groups.

Table 1. ^{1}H (600MHz) and ^{13}C NMR (150MHz) data for compounds **1** and **2**.

Position	1 [a]		2 [a]	
	δ_H, multi (J/Hz)	δ_C (ppm)	δ_H, multi (J/Hz)	δ_C (ppm)
1		166.7		166.4
3		153.7		153.9
4	6.38,s	101.6	6.51,d(1.0)	99.0
4a		137.6		130.7
5		108.7		134.7
6		161.6		159.8
7		128.6	6.50,s	98.6
8		161.4		160.0
8a		99.8		98.3
5-OCH$_3$			3.93,s	61.4
6-OCH$_3$			3.78,s	56.1
6-OH	5.38,s			
7-OH	6.42,s			
8-OH	11.18,s		10.99,s	
9	2.29,s	19.6	2.27,d(1.0)	19.6
10	2.18,s	9.8		

[a] Measured in CDCl$_3$.

The heteronuclear multiple bond correlation (HMBC) correlations (Figure 2) from 8-OH to C-7, C-8 and C-8a, from 7-OH to C-6, C-7, and C-8, from 6-OH to C-5, C-6 and C-7, and from H-10 to C-5, C-6 and C-4a indicated the positions of 6-OH, 7-OH, 8-OH, and CH$_3$-10. In addition, the HMBC correlations from H-9 to C-3 and C-4, and from H-4 to C-5 suggested that CH$_3$-9 was located at C-3. Therefore, the structure of compound **1** was determined as 6, 7, 8-trihydroxy-3, 5-dimethylisocoumarin, as shown in Figure 1.

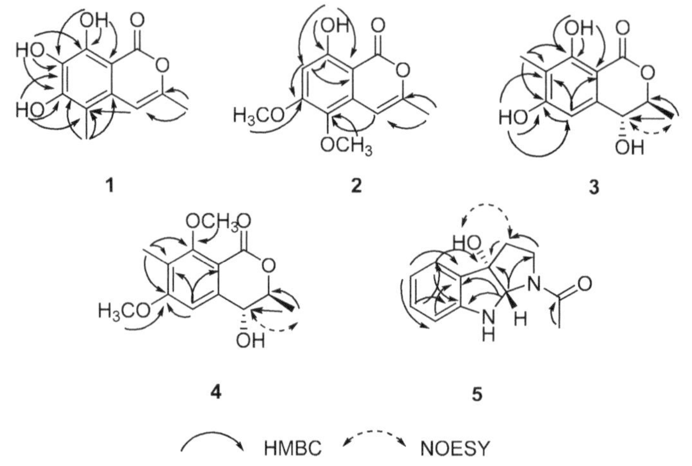

Figure 2. Key HMBC and NOESY correlations for compounds **1–5**.

Compound **2** was obtained as a colorless powder, with the molecular formula C$_{12}$H$_{12}$O$_5$ determined by the HRESIMS at m/z 237.0765 ([M + H]$^+$, calcd. C$_{12}$H$_{13}$O$_5$ 237.0763). The NMR data (Table 1) and HSQC spectrum revealed that compound **2** shared the same isocoumarin skeleton as **1**. However, the CH$_3$ at C-5 in **1** was replaced by a methoxy group in **2**, which was revealed by the HMBC correlations (Figure 2) from the proton of 5-OCH$_3$ to C-5 and from H-4 to C-5. Moreover, the hydroxyl at C-7 in **1** was absent in **2**, and the hydroxyl at C-6 in **1** was changed to a methoxyl in **2**, which was supported by the HMBC correlations from H-7 to C-5, C-6, and C-8a, and from the proton of 6-OCH$_3$ to C-6. Thus, compound **2** was elucidated as 8-hydroxy-5, 6-dimethoxy-3-methylisocoumarin (Figure 1).

The HRESIMS data of compound **3** at m/z 225.0757 ([M + H]$^+$, calcd. C$_{11}$H$_{13}$O$_5$ 225.0757) established the molecular formula C$_{11}$H$_{12}$O$_5$. Its ^{1}H NMR spectrum displayed a singlet for an aromatic methyl group (δ_H 2.09, s), a doublet for another methyl group (δ_H 1.18 s) attached to a methine, two phenolic hydroxyl groups (δ_H 11.27, s and 9.42, s), and a singlet for an aromatic proton (δ_H 6.68, s). The ^{13}C NMR spectrum (Table 2) exhibited the existences of a benzene ring moiety (δ_c 162.7, 155.0, 146.5, 110.7, 101.3, 104.1), an ester carbonyl group (δ_c 171.5), two methyl (δ_c 17.8, 6.9) groups, and two oxygenated methine (δc 84.2, 67.2) groups. These data indicated compound **3** was a dihydroisocoumarin derivative with great similarity to 4, 6-dihydroxymellein [12]. In view of the downfield chemical shift and the peak shape, the hydroxyl group at δ_H 11.27 must be attached to C-8 to form an intramolecular hydrogen bond with the carbonyl (C-1). Hence, the only difference between **3** and **4** is that 6-dihydroxymellein might be an additional methyl at C-7 in **3**. This conclusion was confirmed by the HMBC correlations from CH$_3$-10 to C-6, C-7, and C-8, from 8-OH to C-7, C-8 and C-8a, and from H-5 to C-6, C-7, and C-8a. The deduction that CH$_3$-9 and 4-OH facing *trans* direction was revealed by the strong nuclear Overhauser effect (NOE) correlation between H-4 and CH$_3$-9. The absolute configuration of **3** was concluded to be 3*S*, 4*R*, in that the calculated ECD curve of (3*S*, 4*R*)-**3** had a good agreement with the experimental data (Figure 3).

Table 2. ^{1}H (600MHz) and ^{13}C NMR (150MHz) data for compounds **3** and **4**.

Position	3 [a]		4 [a]	
	δ_H, multi (J/Hz)	δ_C	δ_H, multi (J/Hz)	δ_C
1		171.5		166.5
3	5.33,d(4.8)	84.2	5.30,d(3.0)	82.5
4	4.17,m	67.2	4.29,m	67.2
4a		146.5		149.8
5	6.68,s	101.3	7.00,s	99.8
6		155.0		163.8
7		110.7		119.5
8		162.7		157.1
8a		104.1		110.8
6-OH	9.42,s			
8-OH	11.27,s			
6-OCH$_3$			3.96,s	55.8
8-OCH$_3$			3.99,s	61.2
9	1.18,d(4.8)	17.8	1.22,d(6.6)	18.1
10	2.09,s	6.9	2.11,s	7.8

[a] Measured in (CD$_3$)$_2$CO.

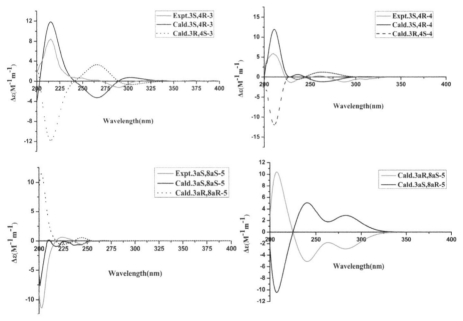

Figure 3. Calculated and experimental ECD spectra for compounds **3**, **4** and **5**.

The NMR features (Table 2) suggested that compound **4** was quite similar to **3**, except for two aromatic hydroxyl groups in **3** being replaced by two methoxy groups (δ_H 3.96, δ_C 55.8, 6-OCH$_3$; δ_H 3.99, δ_C 61.2, 8-OCH$_3$) in **4**, supported by the HMBC correlations from 6-OCH$_3$ to C-6 and from 8-OCH$_3$ to C-8, and by the molecular formula C$_{13}$H$_{16}$O$_5$ according to HRESIMS at m/z 253.1078 ([M + H]$^+$, calcd. C$_{13}$H$_{17}$O$_5$ 253.1076). The positions of 6-OCH$_3$ and 8-OCH$_3$ were further confirmed by HMBC correlations from CH$_3$-10 to C-6, C-7, and C-8, and from H-5 to C-6, C-7, and C-8a. The NOE correlation between H-4 and CH$_3$-9, together with the similar experimental and calculated ECD curves shown in Figure 3, suggested that compound **4** had the same absolute configuration as **3**. Hence, compound **4** was elucidated as (3S, 4R)-4-hydroxy-6, 8-dimethoxy-3, 7-dimethyldihydroisocoumarin.

Compound **5** was obtained as a yellow solid. The molecular formula $C_{12}H_{14}N_2O_2$ (seven degrees of unsaturation) was established from HRESIMS at m/z 219.1127 ([M + H]$^+$, calcd. $C_{12}H_{15}N_2O_2$ 219.1128). The IR spectrum showed an absorption band at 3366 cm^{-1}, corresponding to an amino-group. In addition, the bands at 1615, 1457, and 1413 cm^{-1} were indicative of the amido carbonyl group and aromatic ring system. The ^{13}C NMR (Table 3) and HSQC spectra of compound **5** revealed the presence of an acetyl methyl carbon (δ_C 22.0), one carbonyl carbon (δ_C 170.5), two sp^3 hybridized methene carbons (δ_C 36.4, 47.0), four aromatic methine carbons (δ_C 130.7, 123.4, 119.5, 110.2), one heteroatom substituted methine carbon (δ_C 81.6), one heteroatom attached quaternary carbon (δ_C 86.6), and two phenolic quaternary carbons (δ_C 149.6, 129.2). These data were almost identical to those of the known compound (3a*R*, 8a*S*)-1-acetyl-1, 2, 3, 3a, 8, 8a-hexahydropyrrolo [2,3b] indol-3a-ol isolated from *Streptomyces staurosporeus* fermentation with tryptamine [13], which suggested that compound **5** had the same planar structure as the known compound.

Table 3. ^{1}H (600MHz) and ^{13}C NMR (150MHz) data for compound **5**.

Position	5 ^{a}	
	δ_H, multi (J/Hz)	δ_C
2	3.30,m(10.2,6.6)	47.0
	3.71,m(10.2,3.0)	
3	2.47,m	36.4
	2.54,m	
3a		86.6
3b		129.2
4	7.30,d(7.2)	123.4
5	6.81,dd (7.8,7.2)	119.5
6	7.18,dd(7.8,7.2)	130.7
7	6.64,d(7.8)	110.2
7a		149.6
8	-NH unobserved	
8a	5.32,s	81.6
9		170.5
10	2.03,s	22.0
3a-OH	5.28,s	

^{a} Measured in CDCl$_3$.

The deduction was further confirmed by detailed analysis of the HMBC correlations of **5** (Figure 2). Most chemical shifts of the ^{1}H NMR data of **5** were also consistent with those of the known compound with differences $\leq$ 0.03 ppm in deuterated chloroform. However, the chemical shift of one of the H-3 protons at δ_H 2.70 in the latter was shifted upfield by 0.16 ppm to 2.54 ppm in the former (compound **5**), suggesting that the absolute configuration at C-3a in compound **5** might be different from that of the known compound. Moreover, the specific rotations of the two compounds greatly differed in value (**5**, [α]$_D^{25}$ 23.06; the known compound, [α]$_D^{25}$ 72.00; in MeOH), suggesting that they were a pair of diastereoisomers. Furthermore, NOE correlations between one of the protons of H-3 at δ_H 2.54 and the proton at δ_H 5.28, together with no NOE correlations between H-8a (δ_H 5.32) and the proton at δ_H 5.28, suggested that the proton at δ_H 5.28 should be 3a-OH, which was opposite to H-8a. Thus, the relative configuration of **5** was different from that of the known compound. On the basis of this deduction, the theoretical ECDs were calculated. From Figure 3, the experimental curve of compound **5** matched well with that of the calculative (3a*S*, 8a*S*)-**5** and greatly diverged from those of (3a*R*, 8a*R*)-**5**, (3a*R*, 8a*S*)-**5**, and (3a*S*, 8a*R*)-**5**, suggesting that the absolute configuration of **5** was 3a*S*, 8a*S*. Since 1-acetyl-1, 2, 3, 3a, 8, 8a-hexahydropyrrolo [2,3b] indol-3a-ol had been synthesized as a racemic mixture without specifying the relative configuration [14], it is impossible to judge whether compound **5** was a new compound or not. However, it can be concluded that this is the first report of **5** as a new natural product. Notably, as shown in the HPLC profiles (Figure S34), this new metabolite was produced after adding the ingredient (2*R*, 3*R*)-3, 5, 7-trihydroxyflavanone 3-acetate of the host [10], which exhibits weak antifungal activity against the research fungus L29, suggesting that the pressure to inhibit growth from the antifungal agent of the host changed the metabolic pathway of the fungus. To the best of our knowledge, this is the first example of a new natural product from plant endophytic fungus generated according to such a strategy.

Compounds **1**–**3** and **5** were evaluated in vitro for antifungal activity towards three phytopathogenic fungi: *Fusarium oxysporum* (*F. oxysporum*), *Penicillium italicum* (*P. italicum*), and *Fusarium graminearum* (*F. graminearum*) (Table 4). Compared to the positive control triadimefon, all of the compounds exhibited higher inhibitory activities against *F. oxysporum* and *F. graminearum* except for compound **1**. This compound displayed good activity against *F. oxysporum* (minimum inhibitory concentration (MIC) value = 112.6 μM) and very weak activity towards *F. graminearum* (MIC value = 900.0 μM). The activities towards *P. italicum* for all compounds except compound **3** (inactive, MIC value > 900.0 μM), appeared to range between modest and high values (MIC values from 450.4 μM to 57.3 μM). It should be noted that compound **5** exhibited approximately twelve-, three-, and eighteen-fold stronger activities against *F. oxysporum*, *P. italicum*, and *F. graminearum* than triadimefon, respectively. Moreover, the activities on *F. oxysporum* for compounds **1** and **2**, and on *F. graminearum* for compounds **2** and **3**, were approximately three and two point five times higher than those of triadimefon, respectively. The activities of compound **4** were not assessed because the amount produced was too small.

Table 4. Antifungal activity of compounds **1**, **2**, **3**, and **5** measured as minimum inhibitory concentration (MIC) values.

Compounds	*F. oxysporum*	*P. italicm*	*F. graminearum*
	MIC, μM		
1	112.6	450.4	900.0
2	105.8	211.7	211.7
3	223.0	>900.0	223.0
5	28.6	57.3	28.6
Triadimefon [a]	340.4	170.2	510.7

[a] Positive control towards the test fungi.

3. Materials and Methods

3.1. General Experimental Procedures

Optical rotations were recorded using a P-1020 digital polarimeter (Jasco International Co., Ltd., Tokyo, Japan). Ultraviolet (UV) and IR spectra were obtained using a UV-2550 spectrophotometer (Shimadzu Corporation, Tokyo, Japan) and Nicolet iS10 Fourier transform infrared spectrophotometer (Thermo Electron Corporation, Madison, WI, USA), respectively. CD spectra were collected on Chirascan CD spectrometer (Applied Photophysics Ltd., London, UK). ^{1}H and ^{13}C NMR data were collected on an AVIII 600 MHz NMR spectrometer (Bruker BioSpin GmbH Company, Rheinstetten, Germany). The positive ion mode on a quadrupole-time of flight (Q-TOF) mass spectrometer (Thermo Fisher Scientific Inc., Frankfurt, Germany) was used for the measurement of HRESIMS. Semi-preparative HPLC was performed on a 1260 Infinity Series high performance liquid chromatography system (Agilent Corporation, Santa Clara, CA, USA). Common column and thin layer chromatography (TLC) was performed using a 200−300 mesh and G60, F-254 silica gel (Qingdao Haiyang Chemicals Co., Ltd., Qingdao, China), respectively. For gel filtration chromatography, a Sephadex LH-20 (GE Healthcare, Chicago, IL, USA) was used.

3.2. Fungal and Plant Material

M. bontioides was collected from a mangrove in Leizhou Peninsula, China, in May 2014. The strain L29 was obtained from its leaf and was identified as *Botryosphaeria ramosa* on the basis of molecular analysis methods [15]. These cultures are deposited in the College of Materials and Energy, South China Agricultural University. A BLAST research by NCBI showed that the internal transcribed spacer (ITS) sequence (No. MK370738 in GenBank) of the fungal strain was identical to that of *Botryosphaeria ramosa* CMW26167 (No. NR151841.1).

3.3. Cultivation, Extraction, and Isolation

The strain maintained on potato dextrose agar (PDA) medium at 28 °C for 3 days was put into the liquid medium (2% glucose, 2% peptone) and incubated at 28 °C for about 4 days. After that, the culture (6 mL) was transferred to an Erlenmeyer flask with the rice medium (70 mL H_2O, 50 g rice, 0.25 g crude sea salt) treated either with (2*R*, 3*R*)-3, 5, 7-trihydroxyflavanone 3-acetate (0.25 mM) or without this compound (control), and was incubated for 28 days at room temperature. The flavanone was isolated from *M. bontioides*, the host of the fungus *B. ramosa* L29. It was shown to cause an 18.56% growth inhibition rate at the concentration of 0.25 mM to the mycelium of *B. ramosa* L29 by a petri plate mycelia growth rate method [16]. The media (20 bottles) were continuously extracted three times using 95% ethanol. The solvent was evaporated in vacuo and was extracted three times with ethyl acetate to obtain a dark brown crude extract (25.0 g). The EtOAc extracts of different conditions were analyzed by reversed-phase HPLC (Hypersil BDS C18 column, 150 × 4.6 mm, 5 μm) using a gradient of MeOH/H_2O (20:80–80:20, 0–30 min; 80:20–100: 0, 30–45 min; 100: 0, 45–60 min) at a flow rate of 1.0 mL/min, and recorded at 254 nm. Then, the EtOAc extract was subjected to silica gel column chromatography eluted with a gradient of petroleum ether/EtOAc (90:10, 85:15, 75:25, 50:50, 25: 75, 15:80, *v/v*) to afford six fractions (Fraction 1–6). Fraction 1 was chromatographed on a silica gel column, eluted with a gradient system of petroleum ether/EtOAc (95:5, 90:10, 85:15, *v/v*) to provide Fraction 1.1–1.5; Fraction 1.2 was separated through a Sephadex LH-20 chromatograph (MeOH as eluent) to give Fraction 1.1.1–1.1.2, and Fraction 1.1.1 was purified through semi-preparative HPLC (MeOH/H_2O,90:10, *v/v*; 2.0 mL/min) to yield compound **2** (3 mg, t_R = 33.5 min). Fraction 2 was eluted with a gradient system of petroleum ether/EtOAc (90:10, 85:15, 80:20, 75:25, *v/v*) on a silica gel column to give Fraction 2.1–2.4; Fraction 2.2 was applied on semi-preparative HPLC (MeOH/H_2O, 79:21, *v/v*; 2.0 mL/min) to provide compound **4** (1.5 mg, t_R = 29.3 min). Then, Fraction 3 was fractioned by silica gel column chromatography using petroleum ether/ EtOAc (80:20, 75:25, 70:30, 60:40, 50:50, 40:60, 30:70,20:80, *v/v*) as eluents to afford Fraction 3.1–3.8; Fraction 3.4 was subjected to semi-preparative HPLC (MeOH/H_2O, 68:32, *v/v*; 2.0 mL/min) to yield compounds **3** (2.8 mg, t_R = 22.1 min) and **1** (2.5 mg, t_R = 25.2 min). Further separation of Fraction 3.7 by semi-preparative HPLC (MeOH/H_2O, 65:35, *v/v*; 2.0 mL/min) led to the isolation of compound **5** (2 mg, t_R = 20.4 min).

Compound **1**. Yellow powder; IR (KBr) ν_{max} 3244, 2940, 1682, 1622, 1153, 1164, 1615, 834 cm^{-1}; HRESIMS m/z 223.0608 ([M + H]$^+$, calcd. for $C_{11}H_{11}O_5$ 223.0606); ^{13}C NMR and ^{1}H NMR (Table 1).

Compound **2**. Colorless powder; HRESIMS m/z 237.0765 ([M + H]$^+$, calcd. for $C_{12}H_{13}O_5$ 237.0763); ^{13}C NMR and ^{1}H NMR (Table 1).

Compound **3**. Colorless powder; UV (CH$_3$CN) λ_{max} (log ε) 217 (1.97), 261 (0.68), 298 (0.22) nm; $[\alpha]_D^{25}$ 16.80 (c 0.30, MeOH); HRESIMS m/z 225.0757 ([M + H]$^+$, calcd. for $C_{11}H_{13}O_5$ 225.0757); ^{13}C NMR and ^{1}H NMR (Table 2).

Compound **4**. Colorless powder; UV (CH$_3$CN) λ_{max} (log ε) 217 (0.90), 260 (0.31) nm; $[\alpha]_D^{25}$ 16.00 (c 0.25, MeOH); HRESIMS m/z 253.1078 ([M + H]$^+$, calcd. for $C_{13}H_{17}O_5$ 253.1076); ^{13}C NMR and ^{1}H NMR (Table 2).

Compound **5**. Yellow solid; UV (CH$_3$CN) λ_{max} (log ε) 204 (2.26), 234 (0.97), 297 (0.17) nm; IR (KBr) ν_{max} 3366, 2922, 1615, 1457, 1413, 1313, 1188, 1061, 752 cm^{-1}; $[\alpha]_D^{25}$ 23.06 (c 0.19, MeOH); HRESIMS m/z 219.1127 ([M + H]$^+$, calcd. for $C_{12}H_{15}O_2N_2$ 219.1128); ^{13}C NMR and ^{1}H NMR (Table 3).

3.4. ECD Calculations

Conformational analysis of the enantiomers of compounds **3–5** established by NOESY analyses were carried out via searching with the MMFF94s molecular mechanics force field using

Spartan'10 software (Wavefunction Inc., Irvine, CA, USA) within 10 kcal/mol. The Gaussian 09 program (Gaussian Inc., Wallingford, CT, USA) was used to optimize by density functional theory (DFT) at the 6-31G (d, p) level (methanol as the solvent), and to calculate ECD on 6-311 + G(d, p) level by time-dependent density functional theory (TDDFT). Boltzmann statistics were performed for ECD simulations with a standard deviation of 0.3 eV. The softwares SpecDis 1.64 (University of Wurzburg, Wurzburg, Germany) and OriginPro 8.5 (OriginLab, Ltd., Northampton, MA, USA) were used to generate the ECD curves.

3.5. Antifungal Activity Assay

Three fungi including *F. oxysporum*, *P. italicum*, and *F. graminearum* used for bioassay were acquired from the College of Agriculture, South China Agricultural University. The antifungal effects were examined by the two-fold broth dilution method, as previously reported [17].

4. Conclusions

In this study, four new isocoumarin derivatives, which we named botryospyrones A, B, C, and D, along with a new natural tryptamine derivative, (3a*S*, 8a*S*)-1-acetyl-1, 2, 3, 3a, 8, 8a-hexahydropyrrolo [2,3b] indol-3a-ol, were isolated from an endophytic fungus *B. ramosa* L29, obtained from the leaf of *M. bontioides*. Compound **1** remarkably inhibited *F. oxysporum*, compounds **2** and **3** highly inhibited *F. oxysporum* and *F. graminearum*, and compound **5** significantly inhibited *F. oxysporum*, *P. italicum*, and *F. graminearum*. The compounds were more potent than triadimefon, revealing their potential to be used as new antifungal leads. Compound **5** was produced when using the "UISCH" strategy by adding (2*R*, 3*R*)-3, 5, 7-trihydroxyflavanone 3-acetate, a component of *M. bontioides* with weak growth inhibitory activity towards *B. ramosa* L29, into the medium, suggesting that this strategy could be applied to induce an endophyte to produce new bioactive molecules.

Supplementary Materials: The following are available online at http://www.mdpi.com/1660-3397/17/2/88/s1, Figures S1–S34: the NMR and HRESIMS spectra for all the isolated compounds, UV spectra for compounds **3–5**, IR spectra for compounds **1** and **5**, HPLC profiles for *B. ramosa* L29 EtOAc extract cultured in autoclaved rice medium with/without 0.25 mM (2*R*, 3*R*)-3, 5, 7-trihydroxyflavanone 3-acetate.

Author Contributions: Conceptualization, W.D. and C.L.; methodology, W.D. and C.L.; formal analysis, Z.W., Z.S., W.D., and C.L.; investigation, Z.W., J.C., X.Z., Z.C., T.L., Z.S., W.D., and C.L.; data curation, Z.W.; writing—original draft preparation, Z.W. and W.D.; writing—review and editing, Z.W., W.D., and C.L.; supervision, W.D., Z.S., and C.L.; project administration, W.D. and C.L.; funding acquisition, C.L.

Funding: This research was funded by the Natural Science Foundation of Guangdong Province, grant numbers 2015A030313405 and 2018A030313582; the Science and Technology Project of Guangdong Province, grant number 2016A020222019; and the Science and Technology Project of Guangzhou City, grant number 201707010342.

Conflicts of Interest: The authors declare no conflict of interest.

References

1. Blunt, J.W.; Copp, B.R.; Keyzers, R.A.; Munro, M.H.G.; Prinsep, M.R. Marine natural products. *Nat. Prod. Rep.* **2012**, *29*, 144–222. [CrossRef] [PubMed]
2. Guo, W.Q.; Li, D.; Peng, J.X.; Zhu, T.J.; Gu, Q.Q.; Li, D.H. Penicitols A−C and Penixanacid A from the Mangrove-Derived *Penicillium chrysogenum* HDN11-24. *J. Nat. Prod.* **2015**, *78*, 306–310. [CrossRef] [PubMed]
3. Yu, G.H.; Zhou, Q.L.; Zhu, M.L.; Wang, W.; Zhu, T.J.; Gu, Q.Q.; Li, D.H. Neosartoryadins A and B, fumiquinazoline alkaloids from a mangrove-derived fungus *Neosartorya udagawae* HDN13-313. *Org. Lett.* **2016**, *18*, 244–247. [CrossRef] [PubMed]
4. Huang, S.; Ding, W.; Li, C.; Cox, D.G. Two new cyclopeptides from the co-culture broth of two marine mangrove fungi and their antifungal activity. *Pharmacogn. Mag.* **2014**, *10*, 410–414. [PubMed]
5. Li, X.Z.; Li, C.Y.; Wu, L.X.; Yang, F.B.; Gu, W.X. Chemical constituents from leaves of *Myoporum bontioides*. *Chin. Tradit. Herbal Drugs* **2011**, *42*, 2204–2207.

6. Zhu, X.; Zhou, D.; Liang, F.; Wu, Z.; She, Z.; Li, C. Penochalasin K, a new unusual chaetoglobosin from the mangrove endophytic fungus *Penicillium chrysogenum* V11 and its effective semi-synthesis. *Fitoterapia* **2017**, *123*, 23–28. [CrossRef] [PubMed]

7. Zhu, X.; Zhong, Y.; Xie, Z.; Wu, M.; Hu, Z.; Ding, W.; Li, C. Fusarihexins A and B: Novel cyclic hexadepsipeptides from the mangrove endophytic fungus *Fusarium* sp. R5 with antifungal activities. *Planta Med.* **2018**, *84*, 1355–1362. [CrossRef] [PubMed]

8. Li, W.; Xiong, P.; Zheng, W.; Zhu, X.; She, Z.; Ding, W.; Li, C. Identification and antifungal activity of compounds from the mangrove endophytic fungus *Aspergillus clavatus* R7. *Mar. Drugs* **2017**, *15*, 259. [CrossRef] [PubMed]

9. Huang, S.; Chen, H.; Li, W.; Zhu, X.; Ding, W.; Li, C. Bioactive chaetoglobosins from the mangrove endophytic fungus *Penicillium chrysogenum*. *Mar. Drugs* **2016**, *14*, 172. [CrossRef] [PubMed]

10. Huang, L.L.; Li, J.W.; Ni, C.L.; Gu, W.X.; Li, C.Y. Isolation, Crystal Structure and Inhibitory Activity against *Magnaporthe* Grisea of (2*R*, 3*R*)-3, 5, 7-trihydroxyflavanone 3-acetate from *Myoporum Bontioides* A. Gray. *Chin. J. Struct. Chem.* **2011**, *30*, 1298–1304.

11. Prompanya, C.; Dethoup, T.; Bessa, L.J.; Pinto, M.M.M.; Gales, L.; Costa, P.M.; Silva, A.M.S.; Kijjoa, A. New isocoumarin derivatives and meroterpenoids from the marine sponge-associated fungus *Aspergillus similanensis* sp. nov. KUFA 0013. *Mar. Drugs* **2014**, *12*, 5160–5173. [CrossRef] [PubMed]

12. Takenaka, Y.; Hamada, N.; Tanahashi, T. Aromatic compounds from cultured Lichen mycobionts of three *Graphis* species. *Heterocycles* **2011**, *83*, 2157–2164.

13. Yang, S.W.; Cordell, G.A. Metabolism studies of indole derivatives using a staurosporine producer, *Streptomyces staurosporeus*. *J. Nat. Prod.* **1997**, *60*, 44–48. [CrossRef] [PubMed]

14. Kametani, T.; Kanaya, N.; Ihara, M. Studies on the syntheses of heterocyclic compounds. Part 876. The chiral total synthesis of brevianamide E and deoxybrevianamide E. *J. Chem. Soc. Perkin Trans.* **1981**, *1*, 959–963. [CrossRef]

15. Chen, S.; Liu, Y.; Liu, Z.; Cai, R.; Lu, Y.; Huang, X.; She, Z. Isocoumarins and benzofurans from the mangrove endophytic fungus *Talaromyces amestolkiae* possess α-glucosidase inhibitory and antibacterial activities. *RSC Adv.* **2016**, *6*, 26412–26420. [CrossRef]

16. Mohammedi, Z.; Atik, F. Fungitoxic effect of natural extracts on mycelial growth, spore germination and aflatoxin B1 production of *Aspergillus flavus*. *Aust. J. Crop Sci.* **2013**, *7*, 293–298.

17. Wu, Z.; Xie, Z.; Wu, M.; Li, X.; Li, W.; Ding, W.; She, Z.; Li, C. New Antimicrobial Cyclopentenones from *Nigrospora sphaerica* ZMT05, a Fungus Derived from *Oxya chinensis* Thunber. *J. Agric. Food Chem.* **2018**, *66*, 5368–5372. [CrossRef] [PubMed]

MDPI

St. Alban-Anlage 66

4052 Basel

Switzerland

Tel. +41 61 683 77 34

Fax +41 61 302 89 18

www.mdpi.com

Marine Drugs Editorial Office

E-mail: marinedrugs@mdpi.com

www.mdpi.com/journal/marinedrugs

Lightning Source UK Ltd.
Milton Keynes UK
UKHW050318231222
414274UK00004B/101